Yakov Perelman

Astronomy
for
Fun

For general information on our products and services, please contact us on prodinnova@mail.com

Printed in the United States of America.

ISBN : 978-2917260180

10 9 8 7 6 5 4 3 2 1

Yakov Perelman

Astronomy
for
Fun

Contents

PREFACE

Astronomy is a fortunate science; it needs no embellishments, said the French savant Arago. So fascinating are its achievements that no special effort is needed to attract attention. Nonetheless, the science of the heavens is not only a collection of astonishing revelations and daring theories. Ordinary facts, things that happen day by day, are its substance. Most laymen have, generally speaking, a rather hazy notion of this prosaic aspect of astronomy. They find it of little interest, for it is indeed hard to concentrate on what is always before the eye.

It is this daily aspect of the science of the skies, its beginnings, not later findings, that mainly -but not exclusively- form the contents of *Astronomy for Fun.* The purpose of the book is to initiate the reader into the *basic* facts of astronomy. But do not take it as a primer, since our presentation differs essentially from any text-book. Ordinary facts with which you may be acquainted are couched here in unexpected paradoxes, or slanted from an odd and unexpected angle, solely with a view to excite imagination and quicken interest. We have tried to free the theme as far as possible from the professional "terminology" and technical paraphernalia that so often make the reader shy of books on astronomy.

Books on popular science are often rebuked for not being sufficiently serious. In a way the rebuke is just, and support for it can be found (if one has in mind the exact natural sciences) in the tendency to avoid calculations in any shape or form. And yet the reader can really master his subject only by learning how to reckon, even though in a rudimentary fashion. Hence, both in *Astronomy for Fun* and in other books of this series, the author has not attempted to avoid the simplest of calculations. True, he has taken care to present them in an easy form, well within the reach of all who have studied mathematics at school. It is his conviction that these exercises help not only retain the knowledge acquired; they are also a useful introduction to more serious reading.

The book contains chapters relating to the Earth, the Moon, planets, stars and gravitation. The author has concentrated in the main on materials not usually discussed in works of this nature. Subjects omitted in the present book, will, he hopes, be treated in a second volume. The book, it should be said, makes no attempt to analyze in detail the rich content of modern astronomy.

Yakov Perelman

CHAPTER ONE

THE EARTH, ITS SHAPE AND MOTIONS

The Shortest Way: on Earth and Map

The teacher has chalked two dots on the blackboard. She asks the little boy before her to find the shortest distance between the two points.

A moment's hesitation and the schoolboy carefully draws a curvy line.

"Is that the shortest way," the teacher asks in surprise. "Who taught you that?"

"My Dad. He's a taxi-driver."

The naïve schoolboy's drawing is, of course, a joke. But I suppose you, too, would grin incredulously, were you told that the broken, arched line on Figure 1 was the shortest way from the Cape of Good Hope to the southern tip of Australia! You would be still more amazed to learn that the roundabout way from Japan to the Panama Canal, depicted on Figure 2, is shorter than the straight line between these two places on the same map!

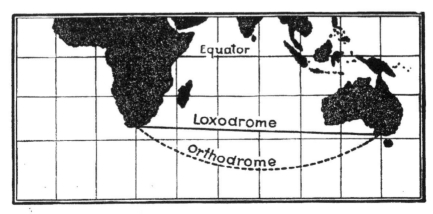

Fig. 1. Nautical charts designate the shortest way from the Cape of Good Hope to the southern tip of Australia not by a straight line ("loxodrome") but by a curve ("orthodrome").

A joke, you might say, but the plain truth, nevertheless, a fact that all cartographers will vouch for.

To make matters clear we ought to say a few words about maps in general and nautical charts in particular. It is no easy matter to draw a part of the Earth's surface, because it is shaped like a ball. Everyone knows that when a sphere is flattened out, there are bound to be creases and rents. Whether we like it or not we have to put up with the inevitable cartographical distortions. Many ways of drawing maps have been devised, but they all have defects of one kind or another.

Seamen use maps charted in the manner of Mercator, the XVI century Flemish cartographer and mathematician. This method is called "Mercator's Projection." The navigator's chart is easily recognized by its network of criss-crossing lines; both the meridians and the latitudes are indicated by parallel straight lines, at right angles (see Figure 5).

Imagine now that your aim is to find the shortest route from one seaport to another, both being on one and the same *parallel*. At sea you can sail in any direction, and, if you know how, you will always be able to find the shortest way. You would naturally think it shortest to travel along the parallel of the two ports -a straight line on our map. After all, what could be shorter than a straight line! But you would be mistaken. The route along the parallel is not the shortest one.

Indeed, on the surface of a ball, the shortest way between two points is the joining arc of the *great* circle.[1] The latitude is a *small* circle, however. The arc of the great circle is less curved than its counterpart on any small circle passing through these two points; the longer radius gives the lesser curve. Take a piece of thread and 'stretch it across the globe between the two points we have chosen (see Figure 3): you will find it does not follow the parallel of latitude at all. Our piece of thread unquestionably points to the shortest route, but, if on the globe it does not coincide with

-1- The great circle on the surface of a sphere is any circle, the centre of which coincides with the centre of the given sphere. All other circles are called small circles.

the parallel of latitude, then on the nautical charts too, where the parallels of latitude are indicated by straight lines, the shortest route will not be a straight line, and thus any line that does not coincide with these straight lines can be only a curved one.

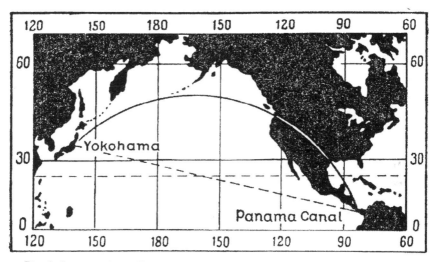

Fig. 2. It seems incredible that the curve linking Yokohama with the Panama Canal is shorter on the nautical chart than the straight line between these two points.

This makes it clear why, on the navigator's chart, the shortest route is not a straight, but curved line.

In choosing a route for the Petersburg-Moscow Railway, the story goes, the engineers could not agree on the site. Tsar Nicholas I got over the difficulty in a "straight line" -he asked for a ruler and drew a straight line between Petersburg and Moscow. Had Mercator's Chart been at hand, the predicament would have been somewhat embarrassing- the railway would have been curved, not straight.

By means of a simple reckoning anyone undeterred by calculations will see for himself that the curved route on the chart is actually shorter than the one we take to be straight. Imagine that our hypothetical seaports are on the same latitude as Leningrad, i.e., the 60th parallel, and that they are 60° apart.

14

Fig. 3. A simple way of finding the really shortest way between two points is to stretch a piece of thread between the given points on a globe.

On Figure 4 the dot *O* designates the centre of the globe, and *AB* the 60° tare of the latitudinal circle where ports *A* and *B* lie. The dot *C* designates the centre of the latitudinal circle. On drawing through the two ports an imaginary *great* circle, arc with its centre at *O* -the centre of the globe- its radius thus being *OB=OA=R*, we shall find it approximating, but not coinciding, with arc *AB*.

We now reckon the length of each arc. As points *A* and *B* are on the 60° latitude, the radii *OA* and *OB* form a 30° angle with *OC* the latter being

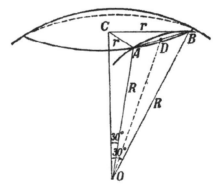

Fig. 4. How to calculate the distances between the points *A* and *B* on a sphere along the arcs of the parallel and the great circle.

the imaginary global axis. In the right-angled triangle ACO, the side AC ($=r$), adjacent to the right angle and opposite the 30° angle, equals half the hypotenuse AO: hence $r=R/2$. As the length of the arc AB is one-sixth the length of the latitudinal circle, which in turn is half as long again as the great circle (the radius being accordingly half less) the length of the small circle arc AB is as follows:

$$AB=1/6 \text{ x } 40{,}000/2 = 3{,}333 \text{ km}$$

To determine the length of the great circle arc between them, we must find the value of angle AOB. As the chord AB, joining the ends of the 60° small circle arc, is the side of an equilateral rectangle inscribed in the same small circle, $AB = r = R/2$. If we draw a straight line OD, joining the point O, the centre of the globe, with the point D halfway along the chord AB, we obtain the right-angled triangle ODA, with the angle D the right angle.

If DA is ½ AB and OA is R, hence the sinus AOD = AD: AO = $R/4$: R = 0.25. We find (from the appropriate tables) that angle AOD is equal to 14°28'30" and hence, the angle AOB is equal to 28°57'.

It will now be easy to find the shortest way, taking the length of one minute of the globe's great circle to be one nautical mile, or about 1.85 km. Hence, 28°57' = 1737' ≈ 3,213 km.

Thus, we have found that the route along the latitudinal circle, indicated on nautical charts by a straight line, is 3,333 km, whereas the great circle route, a curved line on the chart, is 3,213 km, or 120 km shorter.

Equipped with a piece of thread and a school globe, you will easily find our drawings correct and see for yourself that the great circle arcs are actually as shown there. The seemingly "straight" sea-route from Africa to Australia, traced on Figure 1, is 6,020 miles, whereas the "curved" route is only 5,450, or 570 miles (1,050 km) less.

On the navigator's chart the "straight" air line linking London and Shanghai would cut across the Caspian Sea, whereas the shortest way is north of Leningrad. One can well imagine how important this is from the standpoint of saving time and fuel.

Whereas in the era of the sailing vessel time was not always an item of value -man did not then regard "time" as "money"- with the advent of the steamship, every extra ton of coal used meant money. That explains why ships take the shortest course, relying chiefly on charts not of Mercator's Projection, but on what is called the "Central" projection charts indicating the great circle arcs by straight lines.

Why, then, did the seafarers of olden times use such deceptive charts and pick on disadvantageous routes? You would be wrong if you thought that the seamen of old knew nothing of the specific qualities of the navigator's chart we have just mentioned. Naturally, that is not the real reason. The point is that, along with their inconveniences, the charts of Mercator's Projection possess a number of valuable, points for mariners. Firstly, they retain the outline, without distortion, of separate small parts of the globe. This is not altered by the fact that the greater the distance from the equator the more elongated are the contours. In high latitudes the distortion is so great that anyone ignorant of the peculiar features of the navigator's chart would take Greenland to be as large as Africa, or Alaska bigger than Australia, though, actually, Greenland is 15 times smaller than Africa, while Alaska, even together with Greenland, would not be more than half the size of Australia. He would have an absolutely wrong conception of the size of the different continents. But the mariner acquainted with these peculiarities would not be at a disadvantage, because within the small map-sections the navigator's chart provides an accurate picture (Figure 5).

The nautical chart is, moreover, an asset in solving the practical tasks of navigation. It is, in its way, the only chart on which the true straight course of a vessel is indicated by a straight line. To steer a steady course means keeping on in one and the same direction, along one and the same rhumb, or, in

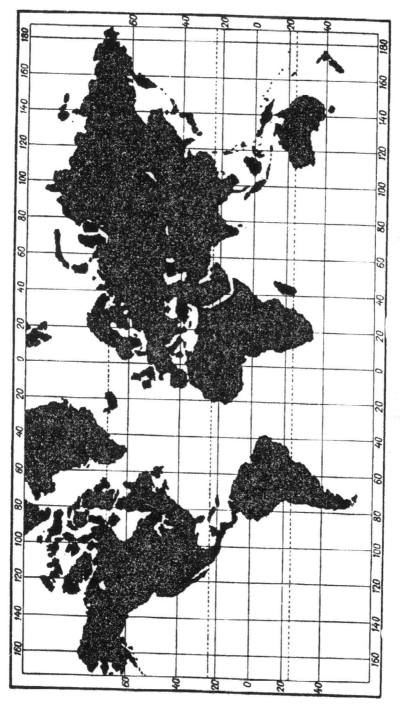

Fig. 5. A nautical or Mercator's chart of the world. These charts strongly dilate the outlines of territories placed far from the equator. Which is bigger: Greenland or Australia? (see text for answer).

Chapter 1

other words, crossing all the meridians at the same angle. This course, known as the loxodrome, can, however, be indicated as a straight line only on a chart where the meridians are straight parallel lines.[2] Since the meridians on the globe are intersected by the latitudes at right angles, this chart should depict the latitudes also as straight lines, perpendicular to the meridians. To put it briefly, we achieve the lattice of coordinating lines that distinguishes the nautical chart.

You will now appreciate why seamen are s3 attracted to Mercator's Projection. To set the course for the port of destination, the navigator joins the points of departure and destination by ruler, and finds the angle between the given line and the meridian. By keeping to this course while at sea, the navigator steers his ship unerringly to his goal. It will be seen, therefore, that while the "loxodrome" is not the shortest or the most economical way, it is, to a degree, a highly convenient course for the seafarer. To reach, say, the southern tip of Australia from the Cape of Good Hope (see Figure 1), the course S 87°50' must the followed undeviatingly. But if we want to get there by the shortest way, along what is known as the orthodrome, we would be forced, as you can see from the picture, to change the course continually, beginning with S 42°50' and ending with N 53°50' (this would be attempting the impossible because our shortest way would take us into the ice-wall of the Antarctic).

The two courses -the "loxodrome" and "orthodrome"- coincide in great circle steering only along the equator or any of the meridians, which are indicated on the nautical chart by a straight line. In all other cases they diverge.

Degree of Longitude and Degree of Latitude

Question

I take it that readers are undoubtedly acquainted with geographical longitude and latitude. But I fear that not all will be

-2- Actually, the loxodrome is a spiral, hugging the globe obliquely.

able to give the correct answer to the following question:

Is it always that a degree of latitude is longer than a degree of longitude?

Most are convinced that each parallel is shorter than the meridian. And since degrees of longitude are measured off the parallels, and those of latitude, off the meridians, the inference is that in no circumstance can the former be longer than the latter. But here they forget that the Earth is not a perfectly round sphere, but an ellipsoid, bulging out slightly at its equator. On this ellipsoid, not only the equator, but even its immediately adjacent parallels are longer than the meridians. According to calculations, roughly up to 5° of latitude, the degrees of the parallels, viz., longitude, are, longer than the degrees of the meridian, viz., latitude.

In What Direction Did Amundsen Fly?

Question

Which direction did Amundsen take when returning from the North Pole, and which on the way back from the South pole?

Give the answer without peeping into the diary of this great explorer.

Answer

The North Pole is the northernmost point of the globe. Whichever way we go from it, we shall always be moving south. In returning from the North Pole, Amundsen could go only south, there being no other direction. Here is an entry from the diary of his North Pole flight aboard the *Norge:*

"The *Norge* circled in the neighborhood of the North Pole. Then we continued on our flight.... We took a southerly course for the first time since our dirigible had left Rome."

By the same token Amundsen could go only *north* when returning from the South Pole. There is a rather hoary anecdote about the Turk who found himself in the "easternmost" country. "East in front, east to the right, east to the left. And what of the west? Perchance, you think it can be spied as a barely visible moving speck in the distance?... You're wrong! East at the back, too. In short, everywhere and all around, nothing but an endless east."

A country facing east on all sides is an impossibility for our Earth. But there is a point with the south all around, just as there is a point hemmed in everywhere, by an "endless" north. At the North Pole it is possible to build a house with all four walls facing south. This is, indeed, a task the Soviet explorers at the North Pole could have actually performed.

Five Ways of Reckoning Time

We are so used to clocks and watches that we fail even to realize the import of their indications. I think I am right in saying that not many readers will know how to explain what they mean when they say:

"It's now 7 p.m."

Is it only that the small hand points to the figure seven? And what does this figure mean? It shows that after midday, so much of the day has passed. But after what midday and, first of all, so much of what day? What is a day? The day, of which the saying "morning, noon and night, a day has taken flight" speaks, is the duration of a complete rotation of our sphere with respect to the Sun. For practical purposes it is measured as follows: two successive passages of the Sun (to be more exact, of its centre) through an imaginary line in the sky connecting the point directly overhead, the "zenith," with the south point of the horizon. The duration varies, with the Sun crossing this line a little earlier or later. It is impossible to set a watch by this "true noon." Even the most skilled craftsman cannot make a watch that will keep time with the Sun; it is too inaccurate. "The Sun shows the wrong time" was the motto of the watchmakers of Paris a century ago.

Our watches are set not to the real Sun but to a fictitious Sun, which neither shines nor warms, but which has been devised for the sole purpose of correctly assessing the time. Imagine a heavenly body whose motion throughout the year is constant, taking exactly the same period of time to go around the Earth as the real Sun seemingly does. In astronomy this fictitious body is known as the "mean sun." The moment of its crossing the zenith-south line is called "mean noon," the interval between two mean noons is known as the "mean solar day," and time thus measured as "mean solar time." Our watches and clocks are set according to this mean solar time. The sundial, however, shows the true solar time for the given location by the Sun's shadow.

The reader might think from what has been said that the globe rotates unevenly on its axis, and that this is the reason for the variation in the length of the true solar day. He would be wrong, however, for this variation is due to the unevenness of another of the Earth's motions in its passage round the Sun. Bear a little and you will see why this affects the length of the day. Turn to Figure 6. Here you see two successive positions of the globe. First the left position. The bottom right arrow shows the direction of the Earth's rotation, viz., counter-clockwise, if viewed from the North Pole. At point A it is now noon; this point is directly opposite, the Sun. Now imagine that the Earth has made one complete rotation; in this time it has shifted right and taken up a second position. The Earth's radius with respect to point A is the same as a day back, but, on the other hand, point A is no longer directly opposite the Sun. It is not noon for anyone at point A; since the Sun is left of the line, the Earth will have to rotate a few minutes more for noon to reach point A.

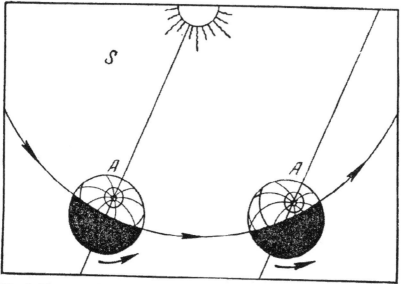

Fig. 6. Why are solar days longer than sidereal days? (see text for details).

What, then, does this imply? That the interval between two true solar noons is longer than the time needed for the Earth to make complete rotation. Were the Earth to travel evenly around the Sun along a circular orbit, with the Sun at the centre, the difference between the real period d rotation and the one we presume with respect to the Sun would be constant day in, day out. This is easily established, especially if we take into consideration the fact that these tiny fractions add up in the course, of one year to make one whole day (in its orbital motion the Earth makes one extra rotation a year); consequently the actual duration of each rotation equals

365.25 days : 366.25 = 23 hours 56 minutes 4 seconds

We might note, incidentally, that the "actual" length of a day is simply the period of the Earth's rotation vis-à-vis any star: hence the term "sidereal" day.

Thus the sidereal day is, on the average, 3 min. 56 sec., or, in round figures, four minutes less than the solar day. The difference is not uniform, firstly, because the Earth's orbit around the Sun is elliptic, not circular, with the Earth moving faster and slower *as* parts of it are nearer and farther from the Sun, and, secondly, because the axis of the Earth's rotation is inclined towards the elliptic. These are the two reasons why on different dates the true and mean solar times vary in terms of minutes, reaching as much as 16 on some days. The two times will coincide only four times a year, viz., April 15, June 14, September 1 and December 24. And conversely, on February 11 and November 2 the difference is greatest -about a quarter of an hour. The curve on Figure 7 shows the degree of discrepancy at different times of the year.

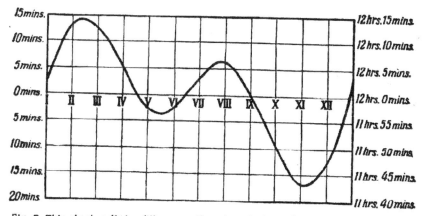

Fig. 7. This chart, called a "time equation chart," shows how great is the discrepancy on any particular day between true and mean solar noon. For instance, on April 1 an accurate clock should show 12:05 at true noon; in other words, the curve gives the mean time at true noon.

Before 1919, people in the U.S.S.R. set their clocks and watches according to local solar time. At each meridian mean noon comes at a different time (the "local" noon), hence each town had its own local time: only train timetables were compiled on the basis of Petrograd time as common for the country. Urban residents recognized two different, "town" and "railway," times, the former of these being the local mean solar time, shown by the town clock, and the latter the Petrograd mean solar time, shown by the station clock. Nowadays railway timetables In the U.S.S.R. are reckoned according to Moscow time.

Since 1919 the counting of time in the U.S.S.R. has been based not on local, but what is called zonal time. The meridians divide the globe Into 24 equal "zones," with every place within the zone having one and the same time, namely, mean solar time, which corresponds to the time of the mean meridian of the particular zone. So nowadays the globe has simultaneously only 24 different times, not the legion that existed before zonal time reckoning was introduced.

To these three ways of reckoning time, viz., (1) true solar time, (2) local mean solar time, and (3) zonal time, we should add a fourth, used only by astronomers, to wit, "sidereal" time, measured on the basis of the above-mentioned sidereal day, which, as we already know, is roughly four minutes less than the mean solar day. On September 22, sidereal and solar time coincide. Thereafter the first jumps four minutes ahead each day.

Finally, there is a fifth way of reckoning time, namely, summer time, observed in the U.S.S.R. all year round, and in most European countries in summer.

Summer time is exactly one hour ahead of zonal time. This is done to save fuel for artificial lighting by starting and ending the workday sooner in the brighter time of year, between spring and autumn. It is achieved by officially setting the hour hand forward. In the West, this is done every spring -at one a. m. the hand is moved to two- while in autumn the hand is reversed.

In the U.S.S.R., clocks have been advanced for the yearly cycle -summer and winter. Although this does not save any more electricity, it ensures more rhythmic working of power plants.

Summer time was first introduced in the Soviet Union in 1917;[3] for a while clocks were advanced two and even three hours.

-3- On the initiative of the author of this book who drafted the appropriate bill.

Yakov Perelman

After a break of several years summer time was again decreed in the U.S.S.R. as from the spring of 1930 and is exactly one hour ahead of zonal time.

Duration of Daylight

For an exact reckoning of daylight duration in any part of the world and on any day of the year, one should refer to the appropriate tables in an astronomical almanac. But the reader will hardly need this pinpoint accuracy; for a comparatively rough-and-ready reckoning the appended drawing will suffice (Figure 8). Its left-hand side indicates the daylight in hours. The lower border gives the Sun's angular distance from the celestial equator, known as the Sun's "declination"; this is measured in degrees. Lastly, the slanting lines correspond to the various latitudes of observation.

To use the drawing we must know the Sun's angular distance ("declination") from the equator to either side for the different days of the year. These figures are tabulated below.

Day of the year	Sun's declination	Day of year	Sun's declination
21 Jan.	-20°	22 July	+20°
8 Feb.	-15	12 Aug.	+15
23 Feb.	-10	28 Aug.	+10
8 March	-5	10 Sept.	+5
21 March	0	23 Sept.	0
4 April	+5	6 Oct.	-5
16 April	+10	20 Oct.	-10
1 May	+15	3 Nov.	-15
21 May	+20	22 Nov.	-20
22 June	+23.5	22 Dec.	-23.5

A few examples of usage:

(1) Find the daylight duration for mid-April, in Leningrad (Lat. 60°).

The table gives us the Sun's declination for mid-April as + 10°, viz., its angular distance from the celestial equator at this particular time. We now find the corresponding figure of 10° on the lower border of our drawing and draw a perpendicular line upwards to intersect the slanting line corresponding to the 60th parallel. Here we turn left to find that the point of intersection corresponds to the figure 14½, which means that the daylight duration we need is roughly 14 hours 30 minutes. We say "roughly" since the drawing does not take into account the effect of what is known as "atmospheric refraction" (see Figure 15).

(2) Find the daylight duration for November 10 in Astrakhan (46° N.L.)

The Sun's declination on Nov. 10 is -17° (it is now in the Southern Hemisphere). Applying the above method we find a duration of 14.5 hours. However, since the declination is now, the figure thus obtained Implies the duration, not of daylight, but of night darkness. So we subtract 14.5 from 24 and get 9.5 hours, the required daylight duration.

We can also reckon the time of sunrise. By halving 9.5 we obtain 4 hours and 45 minutes. From Figure 7 we know that a

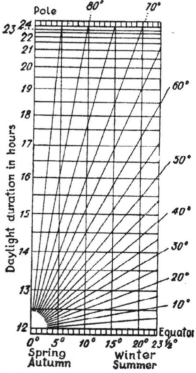

Fig. 8. A daylight duration chart (see text for details).

true noon on November 10, the clock will show 11 hours 43 min. To find the sunrise we subtract 4 hrs. 45 min, and ascertain that the sun will rise at 6 hrs. 58 min. Sunset, on the other hand, will be at 11hrs 43 min + 4 hrs 45 min =16 hrs 48 min, that is, 4 hours 28 min p.m. Thus both drawings (Figures 7 and 8) will, if properly used, substitute for the appropriate tables of an ephemeris.

By using the method just described, you can compile a chart of the rising and setting of the Sun for a whole year for your particular latitude. An example for the 50th parallel, giving also daylight duration, is provided in Figure 9, (compiled, though, on the basis of local, not summer, time). Careful scrutiny will help you draw a similar chart for your own use. Having done so, you will be able, by a cursory glance at your chart, to tell at once the approximate time of sunrise or sunset on any given day.

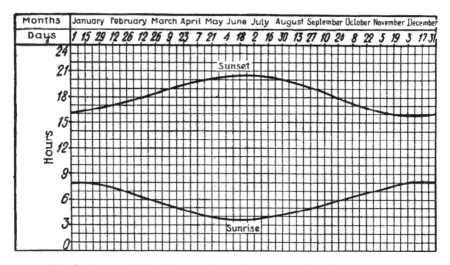

Fig. 9. An annual chart for sunrise and sunset on the 50th parallel.

Extraordinary Shadows

Figure 10 may strike you as being rather queer. The sailor

standing under the full glare of the Sun is practically shadow-less.

Nevertheless, this is a true picture, made not in our latitudes, but at the equator, when the Sun was almost directly overhead, at what is called "zenith."

In our latitudes the Sun is never at zenith, so that a picture such as that described above is out of the question. In our latitudes, when the noonday Sun reaches peak on June 22, it is at zenith everywhere on the northern boundary of the torrid zone (the Tropic of Cancer, i.e., the parallel 23½°

Fig. 10. Almost without a shadow. The drawing reproduces a photo taken near the equator.

N. L.). Six months later, on December 22, it is at zenith everywhere on the parallel 23½° S. L. (the Tropic of Capricorn). Between these boundaries, viz., in the tropics, the noonday Sun is at zenith twice a year, shining in a way that precludes a shadow, or to be more exact, so that the shadow is directly underfoot or underneath.

Figure 11 relates to the Poles. Although, on the contrary, fantastic, it is nonetheless instructive. A man cannot, of course, have shadows in six different places at once. The artist merely wished to convey in a striking fashion the peculiarity of the Polar Sun, which is that the shadow is of one and the same length right around the clock. The reason for this is that at the Poles the Sun does not incline to the horizon throughout the day as it does in our latitudes, but takes a course almost parallel to the horizon. The artist, however, erred in showing too short a shadow compared with the man's height. Were this really so, the Sun would be 40° high, which is impossible for the Poles, where the Sun never rises above 23½°. It can be easily established -the reader with a fair knowledge of trigonometry can make the calculations- that the shortest shadow at the Poles is at least 2.3 times the height of the object casting the shadow.

Fig. 11. At the Pole shadows are of the same length round the clock.

The Problem of the Two Trains

Question

Two absolutely identical trains travelling at the same speed pass each other from opposite directions, one, going west, the other east. Which of the two is heavier?

Answer

The heavier of the two -the one pressing more against the

track- is the train moving contrary to the direction of the Earth's rotation, that is, the westbound train. Moving slower round the Earth's axis, it loses, due to centrifugal effect, less of its weight than the eastbound express.

Fig. 12. The problem of the two trains.

How great is the difference? Let us take two trains moving along the 60th parallel at 72 km/h or 20 m/sec. At this parallel the Earth moves around its axis at a speed of 230 m/sec. Hence the eastbound express has a total circumferential speed of 230 + 20, that is 250 m/sec, and the westbound, a speed of 210 m/sec. The centrifugal acceleration for the first train will be

$$V_1^2/R = 25,000^2/320,000,000 \ cm/sec^2$$

since the radius of the 60th parallel circumference is 3,200 km.

For the second train the centrifugal acceleration is

$$V_2^2/R = 21,000^2/320,000,000 \ cm/sec^2$$

The difference in centrifugal acceleration value between the two trains is

$$(V_1^2 - V_2^2)/R = (25,000^2 - 21,000^2)/320,000,000 \approx 0.6 \ cm/sec^2$$

Since the direction of centrifugal acceleration lies at an angle of 60° to the direction of gravity, we take into consideration only the appropriate fraction of centrifugal acceleration, viz., 0.6 cm/sec² times cos 60° which equals 0.3 cm/sec².

This gives a ratio to gravity acceleration of 0.3/980 or roughly 0.0003.

Consequently the eastbound train is lighter than the westbound by the 0.0003 fraction of its weight. Suppose it consists, say, of 45 loaded boxcars, viz., 3,500 metric tons; the difference in weight would be 3,500 x 0.0003=1,050 kg.

For a ship of 20,000 tons with a speed of 34 km/h (20 knots), the difference would be 3 tons. The decrease in weight in the vessel's eastbound voyage would also be reflected by the barometer; in the above case the mercury would be 0.00015 x 760, or 0.1 mm. lower on the eastbound ship. A Leningrad citizen walking in an easterly direction at a speed of 5 km/h becomes roughly 1½ grams lighter than if he were going in the opposite direction.

The Pocket-Watch as Compass

Most people know how to find their bearings on a sunny day by using a watch. You place the dial so that the hour hand points to the Sun. Then halve the angle formed by this hand and the 6-12 line. The bisector indicates south. It is not difficult to understand why. Whereas the Sun takes 24 hours to traverse its complete path in the heavens, the hour hand moves round the watch face in hail the time, in 12 hours, or dou-

Fig. 13. A simple but inaccurate way of finding the points of the compass with the help of a pocket-watch.

bles the arc in the same time. Hence, if at noon the hour hand indicated the Sun, later it will have outstripped it and doubled the arc. Thus, we have only to bisect this arc to find where the Sun stood at noon, or, in other words, the direction south (Figure 13).

Verification will show this method to be exceedingly crude, being as much as a dozen degrees out at times. To understand why, let us examine the suggested method. The chief reason for the inaccuracy is that the watch, its face upwards, is held parallel to the horizontal plane, whereas the Sun on its daily passage strikes this plane only at the Poles. Elsewhere its path lies at an angle to this plane, up to as much as 90° at the equator. Hence, the watch will give exact bearings only at the Poles; in all other places, greater or lesser deviation is inevitable.

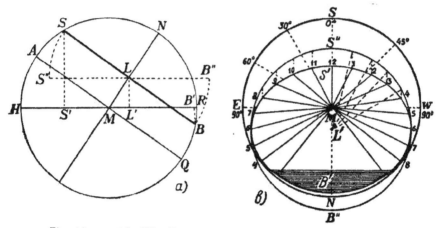

Fig. 14, a and b. Why the watch shows wrong as a compass.

Look at the drawing (Figure 14a). Suppose our observer is standing at M. The point N indicates the Pole, while the circle HASNRBQ -the celestial meridian- passes through the observer's zenith and the Pole. The observer's parallel can be easily ascertained: a protractor measurement of the Pole's altitude above the horizon NR will show it equal to the location's latitude. With his eyes turned H-wards, the observer at M will be facing south. The drawing gives the Sun's daily passage as a straight line -the part above the horizon being day, while the other, below, is night. The straight line AQ indicates the Sun's passage at the equinoxes- when day and night passages are equal. SB, the Sun's passage in summer, is parallel to AQ, but its greater proportion lies above the horizon, and only an insignificant part (recall the short nights of summer) below. The Sun traverses $1/24^{th}$ of the circumference of these circles every hour, or $360°/24$ =$15°$. Nonetheless at three in the afternoon, the Sun will not be exactly SW, as we anticipate ($15° \times 3 = 45°$), the reason for the divergence being that equal arcs of the Sun's passage are not equal in projection on the horizontal plane.

For elucidation, turn to Figure 14b. Here *SWNE* is the horizontal circle as seen from the zenith, and the straight line *SN* the heavenly meridian. *M* is the location of our observer, and *L'* the centre of the circle described by the Sun in its daily passage, as projected on to the horizontal plane. The actual circle of the Sun's path is projected in the form of the ellipse *S'B'*.

Now project hourly divisions of *SB*, the Sun's route, onto the horizontal plane. To do so, turn the circle *SB* parallel to the horizon, to the position S"B", as depicted on Figure 14a. Then divide this circle Into 24 equidistant parts and project the points onto the horizontal plane. New draw from these points of division lines parallel to *SN* to intersect the ellipse S'B', which, if you remember, was the circle of the Sun's passage as projected on the horizontal plane. We clearly perceive the arcs thus obtained to be unequal. To our observer the inequality will seem still greater, located as he is not at point *L'*, the centre of the ellipse, but at point *M*, away from it.

Let us now, for our chosen latitude ($53°$), estimate the degree of inaccuracy in ascertaining the points of the compass by using a watch on a summer day. At this time of the year, the Sun rises between 3 a. m. and 4 a. m. (the boundary of the shaded segment indicating night). The Sun reaches point *E*, east ($90°$), not at 6 a. m. as our watch shows, but at 7:30 a. m. Furthermore, it will reach $60°$ off S, not at 8 a. m. but at 9:30 a.m., and the point $30°$ off S, not at 10 a.m. but at 11 a. m. The Sun will be *SW* ($45°$ to the other side of *S*) not at 3 p.m. but at 1:40 p.m., add will be W not at 6 p.m. but at 4:30 p.m.

Moreover, if we recall that our watch shows Summer Time, which does not coincide with the local true solar time, the inaccuracy will be still greater.

Hence, though the watch can be employed as a compass, it is most unreliable. This make-shift compass will err least at the equinoxes (for thus our observer's location will not be eccentric) and in winter.

"White" Nights and "Black" Days

From mid-April, Leningrad enters a time of "white" nights, the "transparent twilight" and "moonless brilliance," whose fantastic light has engendered so many flights of poetic fancy. Leningrad's "white" nights are so closely associated with literature, that many are prone to think this particular season is the exclusive prerogative of this city. Actually, as an astronomical phenomenon, the "white" nights are true of every point above a definite latitude.

Abstracting ourselves from poetry to astronomic prose, we shall learn that the "white" night is but the mingling of dusk and dawn. Pushkin correctly defined this phenomenon as the meeting of two twilights -morning and evening.

> As tho' to bar the night's intrusion
>
> And keep it out the golden heavens,
>
> Doth twilight hasten to its fusion
>
> With its fellow....

In latitudes where the Sun on its path across the heavens drops a mere $17\frac{1}{2}°$ below the horizon, the sunset is followed almost immediately by dawn, giving the night a bare half hour, even less.

Naturally neither Leningrad nor any other point has a monopoly of this phenomenon. An astronomical survey of the boundary of the "white" nights zone would show it far to the south of Leningrad.

Muscovites, too, can admire their "white" nights -roughly from mid-May till late July. Although not so light as in Leningrad, the "white" night that occurs in Leningrad in May can be observed in Moscow throughout June and in early July.

The southern boundary of the "white" night zone in the U.S.S.R. passes through Poltava, at 49° N. L. (66½ -17½ °), where there is one "white" night a year, namely, on June 22. North of this parallel, the "white" nights are lighter and there are more of them; "white" nights can be observed at Kuibyshev, Kazan, Pskov, Kirov and Yeniseisk. But as all these towns are south of Leningrad, the "white" nights are less (on either side of June 22) and are not so light. On the other hand, at Pudozh, they are lighter than in Leningrad, while in Arkhangelsk, which is close to the land of the unsetting Sun, they are very bright. Stockholm's "white" nights are analogous to those of Leningrad.

When the Sun in its nadir does not dip below the horizon, but just skims it, we have not simply the fusion of sunrise and sunset, but continuous daylight. This is observed north of 65°42', where the domain of the midnight Sun begins. Still farther north, from 67°24', we can also witness continuous night, when dawn and dusk merge at noon, not midnight. This is the "black" day, the "white" night's opposite number, though their brightness is the same. The land of "noonday darkness" is also the land of the midnight Sun, only at a different time of the year. Whereas in June the Sun never sets,[4] in December when the Sun never rises darkness prevails for days on end.

-4- Above Ambarchik Bay, the Sun does not set from May 19 to July 26, and in the vicinity of Tixi Bay from May 12 to August 1.

Daylight and Darkness

The "white" night is clear proof that our childhood notion of equal alternation of night and day on this world of ours is an over-simplification. Actually, the alternation of daylight and darkness is far more variegated and does not fit into the customary pattern of day and night. In this respect the world we live in can be divided into five zones, each with its own alternation of daylight and darkness.

The first zone, outward from the equator in either direction, extends to the 49th parallel. Here, and here alone, is there a full day and a full night in every 24 hours.

The second zone, between 49° and 65½°, embracing the whole of the U.S.S.R. north of Poltava, has continuous twilight around the summer solstice, this being the zone of "white" nights.

Within the third narrow band between 65½° and 67½° the Sun does not set for several days around June 22. This is the land of the midnight Sun.

Characteristic of the fourth zone, between 67½° and 83½°, apart from the continuous day in June, is the long December night, when there are days of no sunrise, and the morning and evening twilight lasts all day. This is the zone of "black" days.

The fifth and last zone, north of 83½°, has a remarkable alternation of daylight and darkness. Here, the break made in the sequence of days and nights by the Leningrad "white" nights, completely upsets the usual order. The six months between the summer and winter solstices, from June 22 to December 22, can, for convenience's sake, be divided into five periods or seasons. First, continuous day; second, alternation of day with midnight twilight, but without proper nights (Leningrad's "white" nights of summer are a feeble imitation of this); third, continuous twilight, with no proper nights or days at all; fourth, continuous twilight alternating with night proper around midnight; and fifth and last, complete darkness all the time. In the next six months, from December to June, these periods follow in reverse order.

On the other side of the equator, in the Southern Hemisphere, the same phenomena are observed, of course, in the corresponding geographical latitudes.

If we have never heard of "white" nights of the "Far South," this is only because the ocean rolls there.

The parallel in the Southern Hemisphere corresponding to Leningrad's latitude does not cross any land at all; there is water everywhere; hence only Polar navigators have had the opportunity of admiring the "white" nights in the south.

The Riddle of the Polar Sun

Question

Polar explorers note a curious feature of the Sun's rays in summer in high latitudes. Although they but feebly heat the Earth's surface there, their effect on all vertical objects, surprisingly enough, is most pronounced.

Steep cliffs and house walls become quite hot, ice-hummocks and the pitch on wooden ships rapidly melt, faces suffer from sunburn, and so on.

What is the explanation?

Answer

This can be explained by a law of physics, according to which the less slanting the rays, the stronger is the effect. Even in summer the Sun in Polar latitudes does not climb very high above the horizon. Beyond the Polar Circle, its altitude cannot exceed half a right angle -in high latitudes it is considerably less.

Taking this as our starting point, we will have no difficulty in establishing that with an upright object the Sun's rays form an angle greater than half a right angle, in other words, they fall steeply on a vertical surface.

This makes it clear why the rays of the Polar Sun, while but feebly heating the surface, intensely heat all upright objects.

When Do the Seasons Begin?

Whether snow is falling, the mercury below zero, or whether the weather is mild, people in the Northern Hemisphere regard March 21 as the end of winter and beginning of spring, that is astronomically. Many really cannot understand why this particular date has been chosen as the dividing line between winter and spring, though, as we have said, a cruel frost may be biting or the weather may be warm and balmy.

The point is that the beginning of the astronomical spring has nothing at all to do with the weather's caprices and vicissitudes. The fact that the beginning of spring is the same for each place in this hemisphere suffices to show that the changes in the weather are of no essential import here. Indeed, the weather cannot be one and the same all over half the world!

In point of fact in fixing the arrival of the seasons, astronomers took not meteorological but astronomical phenomena as their guide, viz., the altitude of the noonday Sun and the ensuing duration of daylight. The weather, then, is but an attending circumstance.

March 21 differs from the other days of the year in that on this date the boundary between light and darkness intersects the two geographical poles. If you hold a globe up to a lamp, you will see that the boundary of the illuminated area follows the meridian, crossing the equator and all the parallels at right angles. Holding the globe thus, turn it on its axis: Every point on its surface will describe a circle, with exactly half in the shade, and the other half in the light. This means that at this particular time of the year, day equals night. This equality is observed all over the world from the North to South Pole. Since the day at this time is 12 hours long, sunrise everywhere is at 6 a.m. and sunset at 6 p.m., naturally, local time.

Thus, the distinguishing feature of March 21 is that all over the world day and night are of equal length on this date. This remarkable phenomenon is known as the "vernal equinox" -vernal because it is not the only equinox. Six months later, September 23 again brings an equal day and night, the "autumnal equinox," ending summer and ushering in autumn. When the Northern Hemisphere has its vernal, the Southern has its autumnal equinox, and vice versa. On one side of the equator winter gives way to spring, on the other, summer yields to autumn. The seasons in the Northern Hemisphere do not tally with those in the Southern Hemisphere.

Let us see how the comparative length of day and night changes throughout the year. Beginning with the autumnal equinox, i.e., September 23, day in the Northern Hemisphere becomes shorter than night. This lasts a full six months, with day at first becoming shorter and shorter until December 22, when it begins to lengthen, and on March 21 it catches up with night. From then on, throughout the other half of the year, day in the Northern Hemisphere is longer than night, lengthening until June 22, and then contracting, but remaining longer than night, until equal length is reached at the autumnal equinox, September 23.

These four dates mark the beginning and the end of the astronomical seasons. For the Northern Hemisphere they are as follows:

March 21 - day equals night - spring begins
June 22 - longest day - summer begins

Sept. 23 - day equals night - autumn begins
Dec. 22 - shortest day - winter begins

Below the equator, in the Southern Hemisphere, spring coincides with our autumn, winter with our summer, and so on.

For the benefit of the reader we suggest at this stage a few questions which, if thought over, will help in assimilating and memorizing what has been said.

1. Where on our planet does day equal night all the year round?

2. At what hour, local time, will the Sun rise in Tashkent on March 21, in Tokyo on the same date, and in Buenos Aires?

3. At what hour, local time, will the Sun set in Novosibirsk on September 23, in New York, and at the Cape of Good Hope?

4. At what hour will the Sun rise at points on the equator on August 2 and February 27?

5. Is it possible to have frost in July and a heat-wave in January?[5]

Three "If's"

Sometimes it is much harder to understand the usual than the unusual. We comprehend the fine points of decimal numeration, which we learn at school, only when we try to use some other system, say of sevens or twelves. Euclid reads like a book only when we probe into non- Euclidean geometry. To really appreciate the role gravity plays in our life, imagine it but a fraction, or on the contrary, a multiple of what it really is, an .artifice we shall resort to later. Meanwhile let us fall back on "if," in order to

-5- Answers: (1) Day and night are always of equal length at the equator, as the boundary between light and darkness also divides the equator into equal halves, irrespective of the Earth's position. (2) and (3) At the equinoxes the Sun rises and sets all over the world at the same hours, 6 am and 6 pm local time. (4) Sunrise at the equator is at 6 am. every day throughout the year. (5) July frosts and January heat-waves are common occurrences in southern latitudes.

better grasp the conditions of the Earth's motion around the Sun.

Let us begin with the axiom, drummed into us during our schooldays, which states that the Earth's axis forms an angle of 66½°, or about ¾ of a right angle, to the Earth's orbital plane. You will appreciate what this means only by imagining this angle to be not three-fourths, but, say, a full right-angle. In other words, suppose the Earth's axis of rotation were perpendicular to its orbital plane, as the Cannon Club in Jules Verne's *Upside Down* dreamed of making it. What changes would this introduce into the tenor of Nature's ways?

If the Earth's Axis Were Perpendicular to the Orbital Plane

Well, suppose that Jules Verne's cannoneers had accomplished their project of "straightening the Earth's axis," and making it form a right angle to the plane of our planet's orbital flight around the Sun. What changes would we observe in Nature?

First of all, the Pole Star -α Ursae Minoris Polaris- would cease to be polar, as the continuation of the Earth's axis would not pass near it, but at some other point around which the celestial dome would revolve.

Further, the alternation of the seasons would be absolutely different or rather there would not be any alternation. What causes the seasons? Why is summer warmer than winter? Let us not evade this commonplace question. At school we obtained but a hazy notion of it, and after school most of us were too busy with other things to bother about it.

Summer in the Northern Hemisphere is warm, firstly, because the tilt of the Earth's axis, the northern end of which is now turned more to the Sun, makes the days longer and the nights shorter. The Sun heats the ground for a longer time and there is no pronounced cooling during the shorter hours of darkness

-the flow of heat increases, the ebb decreases. Secondly, owing again to the inclination of the Earth's axis towards the Sun, the daytime altitude of the latter is high and the rays fall more directly on the Earth. Hence, in summer the Sun sheds more and stronger heat, while the night-time loss is slight. In winter, the reverse is the case the duration of the heat is *shorter* and, moreover, is weaker, while night cooling is pronounced.

In the Southern Hemisphere this process takes place six months later, or earlier, if you wish. In spring and autumn the two poles are equidistant with respect to the Sun's rays; the circle of light almost coincides with the meridians; day practically equals night; and the climate is midway between winter and summer.

What if the Earth's axis were perpendicular to the orbital plane? Would we have this alternation? No, because the globe would always face the Sun's rays from the same angle, and we would have one and the same season at all times of the year. What would this season be? We could call it spring for the temperate and polar zones, though it could, with equal right, be called autumn.

Everywhere and always day would equal night, as is now the case only in the third week of March and September. (This is roughly the case of Jupiter; its axis of rotation is nearly perpendicular to the plane oil its passage around the Sun.)

That would be the case for the temperate zone. In the torrid zone the change in climate would not be so noticeable; for the Poles the contrary would hold. Here due to atmospheric refraction, slightly elevating the Sun above the horizon (Figure 15), instead of setting, it would skim along the horizon. Day, or, to be more exact, early morning, would be perpetual. Although the heat emitted by this low Sun would be slight, it would, since it would shine the year round, make the bleak polar climate appreciably milder. But that would be poor compensation for the damage to the highly developed areas of the globe.

Yakov Perelman

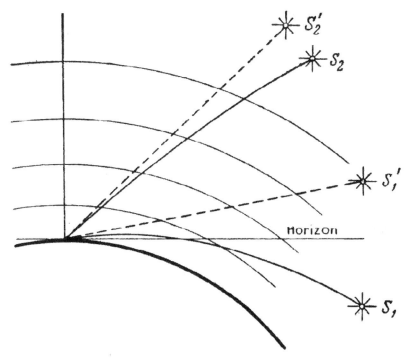

Fig. 15. Atmospheric refraction. The ray from the luminary S_2 is refracted and curved when passing through the layers of the Earth's atmosphere with the result that the observer thinks it is emitted from the point S_2', higher. Although the luminary, S_1 has already sunk below the horizon, the observer still sees it, due to refraction.

If the Earth's Axis Were Tilted 45° to the Orbital Plane

Let us now imagine a 45° tilt of the Earth's axis to the orbital plane. During the equinoxes (around March 21 and September 23) clay would alternate with night as now. In June, however, the Sun would reach zenith at the 45th parallel and not at 23½°; this latitude would become tropical. At the Leningrad latitude (60°) the Sun would be a mere 15° short of zenith, a truly tropical solar altitude! The torrid zone would directly border on the frigid zone, with the temperate zone absent. In Moscow and Kharkov the month of June would be one long, continuous day. In winter, on the contrary, unbroken polar darkness would prevail for weeks, in Moscow, Kiev, Kharkov and Poltava. And the torrid zone in this season would give way to temperate because the noon Sun would not rise higher than 45°.

Naturally, both torrid and temperate zones would lose much by this change. The Polar regions, however, would gain somewhat. Here, after an extremely severe winter, worse than now, there would be a moderately warm summer, when even at the Pole the noonday Sun would be at 45° in the heavens and shine for more than half a year. The Arctic's eternal ice would appreciably retreat under the beneficent action of the sun's rays.

If the Earth's Axis Lay in the Orbital Plane

Our third imaginary experiment is to set the Earth's axis in its orbital plane (Figure 16). The Earth then would revolve around the Sun "in a prone position," rotating on its axis in much the same way as that remote member of our planetary family, Uranus.

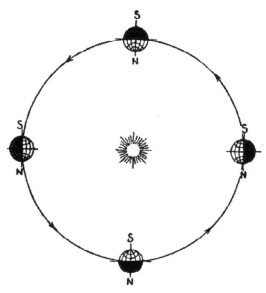

What would happen in this case?

In the vicinity of the Poles there would be a six-month day, during which the Sun would rise spirally from the horizon to zenith, and then descend in the same spiral towards the horizon. This would then give way to a six-month night. The two would be divided by a continu-

Fig. 16. This is how the Earth would move about the Sun if the axis of its rotation were in its orbital plane.

ous twilight of many-day duration. Before disappearing below the horizon, the Sun would traverse the heavens for several days, skimming the horizon. A summer like this would melt all the ice accumulated during the winter.

In middle latitudes the days would quickly become longer with the onset of spring; then, for a period, there would be daylight lasting many days. This long day would set in roughly the number of days coinciding with the number of degrees distant from the Pole, and would last roughly the number of days equal to the degrees of the doubled latitude.

In Leningrad, for instance, this continuous daylight would begin 30 days after March 21, and last 120 days. Nights would reappear 30 days before September 23. In winter the reverse would be the case; continuous daylight would be replaced by continuous darkness of roughly the same duration. Only at the equator would day always equal night.

Uranus' axis is inclined to its orbital plane roughly as described above; its tilt towards the plane of its passage round the Sun is only 8°. One might say of Uranus that it revolves around the Sun *"lying on its side."*

These three "if's" will, in all probability, give the reader a better idea of the relation between climate and the tilt of the Earth's axis. It is not accidental that in Greek the word "climate" means "inclination."

One More "If"

Let us now turn to another aspect of our planet's motions, viz., the form of its orbit. Like every planet, the Earth abides by Kepler's first law, which is that each planet follows an elliptical path of which the Sun is one of the foci.

What is the ellipse of the Earth's path like? Does it differ greatly from a circle?

Text-books and tracts on elementary astronomy often, depict the globe's orbit as a rather strongly extended ellipse. This picture, wrongly understood, is fixed in many minds for life; many people remain convinced that the Earth's orbit is an appreciably elongated ellipse. However, this is not at all so; the difference between the Earth's orbit and a circle is so negligible that it cannot be drawn otherwise than as a circle. Suppose in our drawing the orbit's diameter is one meter. The difference between it and a circle would be less than the thickness of the line drawn to depict it. Even the painter's fastidious eye would fail to distinguish between this ellipse and a circle.

Let us dwell for a moment on elliptical geometry. In the ellipse on Figure 17, AB is its "major axis," and CD, its "minor axis." Apart from the "centre," O, each ellipse has still another two remarkable points, the "foci," set symmetrically on the major axis on either side of the centre. The foci are found in the following manner (Figure 18). A pair of compass legs are stretched to cover a distance equal to the major semi-axis OB. With one leg at C, the end of the minor axis, we describe with the other an arc intersecting the major axis. The points of intersection, F and F, are

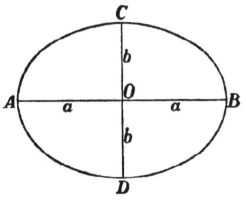

Fig. 17. An ellipse and its axes, major (AB) and minor (CD). Point O designates its centre.

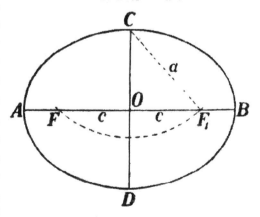

Fig. 18. How the foci of an ellipse are found.

the foci of the ellipse. The equal distances OF and OF_I will now be designated c, and the axes, both major and minor, $2a$ and $2b$. The length c, measured off the length a of the major semi-axis, i.e., the fraction c/a, is the measure of the elongation of the ellipse and is called the "eccentricity." The greater the difference between ellipse and circle, the greater the eccentricity.

We shall have an exact idea of the form of the globe's orbit when we learn the value of its eccentricity. This can be ascertained even without measuring the value of the orbit. The Sun, set at one of the orbit's foci, seems to us on Earth to be of a different size owing to the different distances of the points of the orbit from this focus. At times the visible dimensions of the Sun increase, at times diminish, their ratio exactly conforming to the ratio of the distances between Earth and Sun at the moments of observation. Assume the Sun to be at focus F, of our ellipse (Figure 18). The Earth will be at point A of the orbit about July 1, when we shall see the Sun's smallest disc, its angular value being 31'28". The Earth reaches point B about January 1, when seemingly, the Sun's disc is at its greatest angle-32'32". We now set the following ratio:

$$31'28''/32'32'' = BF/AF_I = (a\text{-}c)/(a\text{+}c)$$

from which get the so-called derivative ratio:

$$a\text{-}c\text{-}(a\text{+}c)/a\text{+}c\text{+}(a\text{-}c) = 31'28''\text{-}32'32''/32'32''\text{+}31'28''$$

or:

$$64''/64' = c/a$$

This means:

$$c/a = 1/60 = 0.017,$$

i.e., the eccentricity of the Earth's orbit is 0.017. All that is needed, therefore, is to take a careful measurement of the Sun's visible disc to determine the form of the Earth's orbit.

Now let us prove that the Earth's orbit differs very little from a circle. Imagine an enormous drawing with the orbit's major semi-axis equal to one meter. What will be the length of the other, minor axis of the ellipse? From the right angled triangle OCF_1 (Figure 18) we find

$$c^2 = a^2 - b^2, \text{ or } c^2/a^2 = (a^2 - b^2)/a^2$$

but c/a is the eccentricity of the Earth's orbit, i.e. $1/60$, We replace the algebraic expression $a^2 - b^2$ by $(a-b)(a+b)$, and $(a+b)$ by $2a$, since b differs but slightly from a. Thus we obtain

$$1/60^2 = 2a(a-b)/a^2 = 2(a-b)/a$$

and hence $(a - b) = a/2 \times 60^2 = 1000/7200$, i.e., less than $1/7$ mm.

We have found that even on this big scale the difference between the length of the major and minor semi-axes of the Earth's orbit is not more than $1/7$ mm. -thinner than a slender penciled line. So we shall not be far wrong if we draw the Earth's orbit as a circle.

But where would the Sun fit in our scheme? In order to place it at the orbit's focus, how far should it be from the centre? In other words, what is the length of OF or OF_1 on our imaginary drawing? The reckoning is rather simple:

$$c/a = 1/60, c = a/60 = 100/60 = 1.7 \text{ cm}$$

On our drawing the Sun's centre should be 1.7 cm. away from the centre of the orbit. But as the Sun itself should be depicted by a circle 1 cm. in diameter, only the practiced eye of the painter would discern that it is not in the centre of the circle.

The practical conclusion is that we can depict the Earth's orbit as a circle, placing the Sun slightly to the side, of the center.

Yakov Perelman

Can this negligible asymmetrical position of the Sun influence the Earth's climate? To discover the likely effect, let us conduct another imaginary experiment, again playing at "if." Suppose the eccentricity of the Earth's orbit were bigger, say, 0.5. Here the focus of the ellipse would divide its semi-axis in half; this ellipse would look roughly like an egg. None of the orbits of the major planets in the solar system have this eccentricity; Pluto's orbit, the most extended, has an eccentricity of 0.25. (Asteroids and comets, however, move along more extended ellipses.)

If the Earth's Path Were More Extended

Imagine the Earth's orbit noticeably elongated, with the focus dividing its major semi-axis into half. Figure 19 depicts this orbit. The Earth, as hitherto, would be at point A, nearest to the Sun, on January 1, and at point B, farthest away, on July 1. Since FB is thrice FA, the Sun would be three times nearer to us in January than in July. Its January

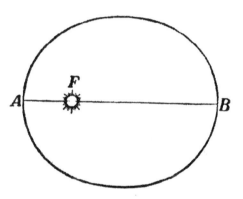

Fig. 19. This is the shape the Earth's orbit would have, if its eccentricity were 0.5. The Sun is at the focus F.

diameter would be triple the July diameter, and the amount of heat emitted would be nine times greater than in July (the reverse ratio of the squared length). What, then, would be left of our northern winter? Only the Sun would be low in the heavens, the days would be shorter and the nights would be longer. But, there would be no cold weather -the Sun's proximity would compensate for the daylight deficit.

To this we must add another circumstance, stemming from Kepler's second law, which is that the radius-vector sweeps over equal areas in equal times.

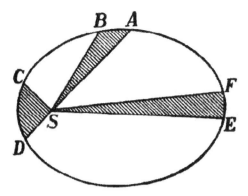

Fig. 20. An illustration to Kepler's second law: if the planet travels along the arcs *AB*, *CD* and *EF* in equal times, the shaded segments must be equal in area.

The "radius-vector" of an orbit is the straight line joining the Sun with the planet, the Earth in our case. The Earth moves along its orbit together with its radius-vector, with the latter sweeping over a certain area; we know from Kepler's law that the sections of the area of an ellipse swept over in equal times are equal. At points nearer to the Sun the Earth should move faster along its orbit than at points farther away, otherwise the area swept over by a shorter radius-vector would not equal the area covered by a longer one (Figure 20).

Applying this to our imaginary orbit we deduce that between December and February, when the Earth is much nearer to the Sun, it moves faster along its orbit than between June and August. In other words, the northern winter is of short duration, whereas summer, on the contrary, is long, as if compensating for the niggardly warmth exuded by the Sun.

Figure 21 furnishes a more exact idea of the duration of the seasons under our imagined conditions. The ellipse depicts the form of the Earth's new orbit, with an eccentricity 0.5. The figures 1-12 divide the Earth's path into the sections which it traverses at equal intervals; according to Kepler's law the sections of the ellipse divided by the radius-vectors are equal in area. The Earth will reach point *1* on January 1, point 2 on Febru-

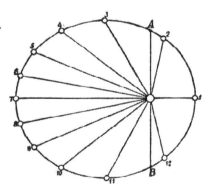

Fig. 21. This is how the Earth would revolve around the Sun, if its orbit were a strongly extended ellipse. (The planet covers the distance between each figure-designated point in equal times—one month.)

ary 1, point 3 on March 1, and so on. The drawing shows that on this orbit the vernal equinox (A) should set in at the beginning of February, the autumnal (B) at the end of November. Thus for the Northern Hemisphere winter would be little more than two months, from the end of November to the beginning of February. On the other hand the season of long days and a high noonday Sun, lasting from the vernal to the autumnal equinox, would be more than 9.5 months.

The reverse would be true of the Southern Hemisphere. The Sun would hang low and short days occur, when the Earth would be farther away from the diurnal Sun and the latter's heat would dwindle to a $1/9^{th}$, conversely, high solar altitude and long days would coincide with a 9-fold increase in the Sun's warmth. Winter would be much more rigorous and far longer than in the North. On the other hand, summer, though short, would be unbearably hot.

Another consequence of our "if." In January the Earth's rapid orbital motion would make the moments of mean and true noon diverge considerably -a difference of several hours. This would make it very inconvenient to follow the mean solar time we now observe.

We now have an idea of the effects of the Sun's eccentric position in the Earth's orbit. First, winter in the Northern Hemisphere should be shorter and milder, and summer longer than in the Southern Hemisphere. Is this really so? Unquestionably, yes. In January the Earth is nearer to the Sun than it is in July by $2 \times 1/60$, i.e., by $1/30$. Hence, the amount of heat received increases $(61/59)^2$ times, i.e., 6 per cent. This somewhat alleviates the severity of the northern winter. Furthermore, the northern autumn and winter together are roughly eight days shorter than the southern seasons; while 4 summer and spring in the Northern Hemisphere are eight days longer than in the Southern Hemisphere. This, possibly, may be the reason for the thicker ice at the South Pole. Below is a table showing the exact length of the seasons in the Northern and Southern Hemispheres:

Northern Hemisphere	Length	Southern Hemisphere
Spring	92 days 19 hrs	Autumn
Summer	93 days 15 hrs	Winter
Autumn	89 days19 hrs	Spring
Winter	89 days 0 hrs	Summer

As you see, the northern summer is 4.6 days longer than winter, the spring 3.0 days longer than autumn.

The Northern Hemisphere will not retain this advantage eternally. The major axis of the Earth's orbit is gradually shifting in space, with the result that the points along the orbit nearest and farthest from the Sun are transferred elsewhere. These motions make one full cycle in 21,000 years. It has been calculated that around 10700 A.D. the Southern Hemisphere will enjoy the above-mentioned advantages of the Northern Hemisphere.

Nor is the eccentricity of the Earth's orbit rigidly fixed; it slowly vacillates throughout the ages between almost zero (0.003), when the orbit is almost a circle, and 0.077, when the orbit is most elongated, resembling that of Mars. Currently its eccentricity is on the wane; it will diminish for another 24 millenniums to 0.003, and will then reverse the process for 40 millenniums. These changes are so slow that their importance is purely theoretical.

When Are We Nearer to the Sun, Noon or Evening?

Were the Earth to follow a strictly circular orbit with the Sun at its central point, the answer would be very simple. We would be nearer to the Sun at noon, when the corresponding points on the surface of the globe, owing to the Earth's axial rotation, are in conjunction with the Sun. The greatest length of this proximity to the Sun would be, for points on the equator, 6,400 km., the length of the Earth's radius.

But the Earth's orbit is an ellipse with the Sun at its focus (Figure 22). Consequently, at times the Earth is nearer to the Sun and at times farther away. For the six months between January 1 and July 1, the Earth moves away from the Sun, during the other six it approaches the Sun. The

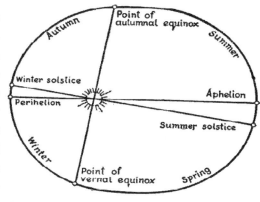

Fig. 22. A diagram of the Earth's passage round the Sun.

difference between the greatest and the least distance is 2 x 1/60 x 150,000,000, i.e., 5,000,000 kilometers.

This variation in distance averages some 28,000 km. a day. Consequently, between noon and sunset (a fourth of a day) the distance from the diurnal Sun changes on the average by 7,500 km., that is, more than from the Earth's axial rotation. Hence, the answer: between January and July we are nearer to the Sun at noon, and between July and January we are nearer in the evening.

Add a Meter

Question

The Earth revolves around the Sun at a distance of 150,000,000 km. Suppose we add one meter to this distance. How much longer would be the Earth's path around the Sun and how much longer the year, provided the velocity of the Earth's orbital motion remained the same (see Figure 23)?

Answer

Now one meter is not much of a distance, but, bearing in mind the enormous length of the Earth's orbit, one might think that the addition of this insignificant distance would noticeably increase the orbital length and hence the duration of the year.

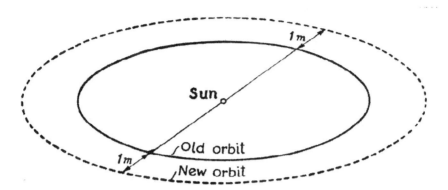

Fig. 23. How much longer would the Earth's orbit be, if our planet were 1 m. further from the Sun? (see text for answer).

However, the result, after totting up, is so infinitesimal that we are inclined to doubt our calculations. But there is no need to be surprised; the difference is really very small. The difference in the length of two concentric circumferences depends not on the value of their radii, but on the difference between them. For two circumferences described on a floor the result would be exactly the same as for two cosmic circumferences, provided the difference in radii was one meter in both cases. A calculation will show this to be so. If the radius of the Earth's orbit (accepted as a circle) is R meters, its length will be $2\pi R$. If we make the radius 1 meter longer, the length of the new orbit will be

$$2\pi(R+1) = 2\pi R + 2\pi.$$

The addition to the orbit is, therefore, only 2π, viz., 6.28 meters, and does not depend on the length of the radius.

Hence the Earth's passage around the Sun, were it set 1 meter more away, would be only 6¼ meters longer. The practical effect of this on the length of the year would be nil, as the Earth's orbital velocity is 30,000 m/sec. The year would be only 1/5000th of a second longer, which we, of course, would never notice.

From Different Points of View

Whenever you drop something, you observe that it falls vertically. You would think it queer if someone else observed it falling not in a straight line. And yet this would be true of any observer not involved together with us in the Earth's motions.

Fig. 24. Anyone on our planet would see a freely falling body drop along a straight line.

Imagine ourselves watching a falling body through the eyes of this observer. Figure 24 shows a heavy ball freely dropping from a height of 500 meters. As it falls, it naturally participates simultaneously in all terrestrial motions. The only reason why we fail to notice these supplementary and far more rapid motions' of the falling body is because we ourselves are involved in them. If we could abstract ourselves from participation in but one of the motions of our planet, we would see the same body falling not vertically but along another path altogether.

Suppose we are watching the falling body from the surface not of the Earth but of the Moon. Although the Moon accompanies the Earth in the latter's revolution around the Sun, it is not involved in its axial rotation. So from the Moon we would see the body make two motions, one vertically downwards, the other, which we had not observed before, at an eastward tangent towards the Earth's surface. The two simultaneous motions add up of course, in accordance with the rules of mechanics, and, as one is uneven and the other even, the resulting motion will occur along a curve. Figure 25 shows this curve, or how a sharp-eyed man in the Moon would see a body falling on the Earth.

Let us go one step further and imagine ourselves on the Sun observing through an extra-powerful telescope the earthward flight of this heavy ball. On the Sun we shall be outside both the Earth's axial rotation and its orbital revolution. Hence, we shall see simultaneously three motions of the falling body (Figure 26): (1) a vertical drop onto the Earth's surface, (2) a motion eastward along a tangent towards the Earth's surface, and (3) a motion round the Sun.

Motion No. 1 covers 0.5 km. Motion No. 2, in the 10 seconds of the body's downward flight, would, at Moscow's latitude, be 0.3 x 10 = 3

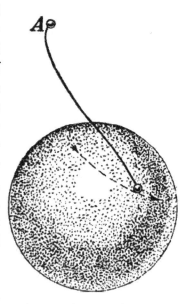

Fig. 25. The man in the Moon would see the same flight as a curve.

Yakov Perelman

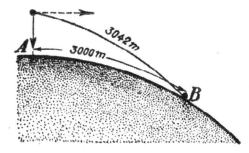

Fig. 26. A body falling freely onto our Earth simultaneously moves at a tangent to the circular route, described by the points of Earth's surface due to rotation.

km. The third, land fastest, motion is 30 km/sec, so that in the 10 seconds of its downward fall it would travel 300 km along the Earth's orbit. Compared with this pronounced shift, the others, the 0.5 km down and the 3 km along the tangent, would hardly be distinguished; from our vantage point on the Sun the eye would be caught only by the main flight. What do we get? Roughly what we see (the appropriate scale is not observed here) in Figure 27. The Earth shifts leftwards, while the falling body drops from a point above the Earth in the right position, to a corresponding point (just slightly lower) on the Earth in the left position. As I said above the correct scale is not observed here -in the 10 seconds the centre of the Earth would shift not 14,000 km, as our artist has made it do for the sake of clarity, but only 300 km.

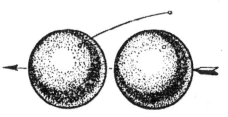

Fig. 27. What anyone observing the falling body shown on Fig. 24 would see from a vantage point on the Sun (scale has been disregarded).

Let us make yet another step and imagine ourselves on a star, i.e., on a remote Sun, beyond the motions even of our own Sun. From there we would observe, apart from the three motions examined, a fourth motion of the falling body with respect to the star on which we are now standing. The value and direction of the fourth motion depend upon the star we have chosen, i.e., on the motion made by the entire solar system with respect to this star.

Figure 28 is a likely case when the solar system moves with respect to the chosen star at an acute angle to the ecliptic, at a velocity of 100 km/sec (Stars have velocities of this order.) In 10 seconds this motion would shift the falling body 1,000 km and, naturally, coin- plicate its flight. Observation from another star would give this path another value and another direction.

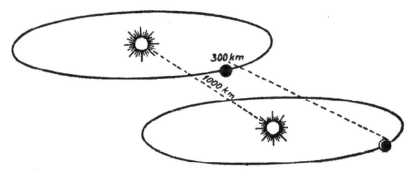

Fig. 28. How an observer on a distant star would see a body falling onto the Earth.

We could go farther still and imagine what the earth wise flight of a falling body would look like to an observer beyond the Milky Way, who would not be involved in the rapid motion of our stellar system with respect to other islands of the universe. But there is no point in doing so. Readers will know by now that, observed from different vantage points, the flight of one and the same falling body will be seen differently.

Unearthly Time

You have worked one hour and then rested for an hour. Are these two times equal? Unquestionably yes, if reckoned by a good timepiece, most people would say. But what timepiece should we use? Naturally, that checked by astronomical observation, or in other words, the one that chimes with the motion of a globe rotating with ideal evenness, turning at equal angles in absolutely equal times.

But how, you may ask, do we know that the Earth's rotation is even? Why are we certain that the two consecutive axial rotations of our planet take equal times? We cannot verify this while the Earth's rotation is itself a gauge of time.

Lately astronomers have found it useful in some cases provisionally to replace this long established model of even motion by another. Here are the reasons and the consequences of this step.

Careful study revealed that in their motions some of the heavenly bodies did not conform to theoretical suppositions, and that the divergence could not be explained by the laws of celestial mechanics. It was found that the Moon, Jupiter's satellites I and II, Mercury, and even the *visual* annual motions of the Sun, i.e., the motion of our own planet along its own orbit, had variations, for which there was no apparent reason. The Moon, for example, swerves from its theoretical path to as much as $1/6^{th}$ of a minute of an arc in some epochs, and the Sun to as much as a second of an arc. An analysis of these incongruities disclosed a feature common to all: at one period these motions gather speed and, subsequently, slow down. Naturally it was inferred that these deviations had one common cause.

Was not this due to the "inaccuracy" of our natural clock, to the unlucky choice of the Earth's rotation as a model of even motion?

The question of replacing the "Earthly clock" was raised: it was provisionally discarded, and the investigated motion measured by another natural timepiece based on the motions either of one or other of Jupiter's satellites, the Moon, or Mercury. This action immediately introduced satisfactory order into the motion of the above-mentioned celestial bodies. On the other hand, the Earth's rotation as measured by this new timepiece was found to be uneven -slowing down for a few dozen years, gathering speed in the next few dozen, and then slowing down once more.

In 1897 the day was 0.0035 seconds longer than in earlier years and in 1918 the same amount less than between 1897 and 1918. The day how is roughly 0.002 seconds longer than it was a hundred years ago.

In this sense we can say that our planet rotates unevenly with respect to other of its motions and also with respect to the motions in our planetary system conventionally accepted as even motions. The value of the Earth's deviations from a strictly even motion (in the sense indicated) is exceedingly negligible: during the hundred years between 1680 and 1780 the Earth rotated slower, the days were longer and our planet accumulated some 30 seconds difference between its "own" and "other" time; then, up to the middle of the 19th century the days shortened and about 10 seconds were sliced off; by the beginning of the present century another 20 seconds had been lost; however, in the first quarter of the 20th century the Earth's motion again slowed down, the days lengthened and a difference of nearly half a minute again accumulated (Figure 29).

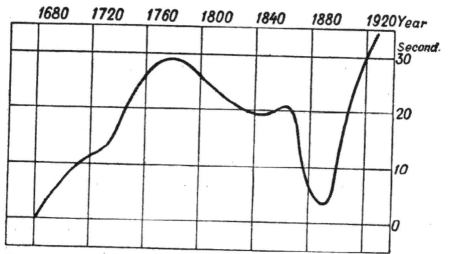

Fig. 29. The curved line shows how far the Earth swerved from even motion between 1680 and 1920. If the Earth rotated evenly this motion would be sketched on the chart as a horizontal line. The ups show a longer day when the Earth's rotation slowed down, and the downs a shorter day when rotation began to gather speed.

Various reasons have been adduced for the changes, for instance, lunar tides, the changes in the Earth's diameter[6] and so on.

-6- The change in the length of the Earth's diameter may escape direct measurement as its accuracy is known only to 100 meters. It would be enough for the Earth's diameter to become a few meters longer or shorter to cause the above-mentioned changes in the day's duration.

It is quite possible that all-round study of this phenomenon will yield important discoveries.

Where Do the Months and Years Begin?

Midnight has struck in Moscow, ushering in the New Year. West of Moscow it is still December 31, while eastwards it is already January 1. However, on our spherical Earth, East and West must inevitably meet this means that there must be a boundary line somewhere dividing the 1st from 31st, January from December and the New Year from the old.

This is called the International Date Line. It passes through the Bering Straits, through the Pacific Ocean, roughly along the meridian 180°. It has been exactly defined by international agreement.

It is along this imaginary line, intersecting the wastes of the Pacific, that the days, the months and the years change for the first time on the globe. Here lies what may be called the threshold of our calendar; it is from this point that every day of the month starts in. This is the cradle of the New Year. Each day of the month appears earlier here than anywhere else; from here it spreads west, circumnavigates the globe and again returns to its birth-place to vanish.

The U.S.S.R. leads the world as host to the new day of the month. At Cape Dezhnev, the day newly born in the waters in the Bering Straits is welcomed into the world and begins its march across every part of the globe. And it is here, at the eastern tip of Soviet Asia, that day ends, after doing service for 24 hours.

Thus, the days change on the International Date Line. The mariners who first circumnavigated the world (before this line was established) miscalculated the days. Here is a true story told by Antonio Pigafetta, who accompanied Magellan on his voyage around the world.

"On July 19, *Wednesday*, we sighted the Cape Verde Islands and dropped anchor... Anxious to know whether our log books were correct we inquired the day of the week. We were told it was Thursday. This surprised us, because our log indicated Wednesday. It seemed unlikely that all of us had made the same mistake of one day...

"We learned later that we had made no mistake at all in our reckoning. Sailing continuously westwards, we had followed the Sun in its path and upon returning to our point of departure should have gained 24 hours upon those left behind. One need only think over this to agree."

What does the seafarer do now when he crosses the date line? To avoid error, he "loses" a day when sailing from east to west, and "adds" a day, when returning. Therefore the story told by Jules Verne in his *Around the World in Eighty Days* about the voyager, who having sailed round the world "returned" on Sunday when it was still Saturday, could not have happened. This could have occurred only in Magellan's times when there was no date line agreement. Equally unthinkable in our time is the adventure described by Edgar Allan Poe in his *Three Sundays in a Week*, about the sailor who after going round the world from east to west met, upon returning home, another who had done the journey in the reverse direction. One claimed that the day before had been Sunday, the other was convinced that the morrow would be Sunday, while their land-lubber friend insisted that today was Sunday.

So as not to fall out with the calendar in a round-the-world trip one should, when travelling east, tarry a little in reckoning the days, letting the Sun catch up, or in other words, count one and the same day twice; on the other hand, when travelling west, he should, on the contrary, lose a day, so as not to lag behind the Sun.

Although this is commonplace, even in our days, four centuries after Magellan's voyage, not everybody is aware of it.

How Many Fridays Are There in February?

Question

What is the greatest and least number of Fridays in February?

Answer

The usual answer is that the greatest number is five -the least, four. Without question, it is true that if in a leap year February 1 falls on a Friday, the 29th will also be Friday, giving five Fridays altogether.

However, it is possible to reckon double the number of Fridays in the month of February alone. Imagine a ship plying between Siberia and Alaska and leaving the Asiatic shore regularly every Friday. How many Fridays will its skipper count in a leap-year February of which the 1st is a Friday? Since he crosses the date line from west to east and does so on a Friday, he will reckon two Fridays every week, thus adding up to 10 Fridays in all. On the contrary, the skipper of a ship leaving Alaska every Thursday and heading for Siberia will "lose" Friday in his day reckoning, with the result that he won't have a single Friday in the whole month.

So the correct answer is that the greatest number of possible Fridays in February is 10, and the least -nil.

CHAPTER TWO

THE MOON AND ITS MOTIONS

New or Old Moon?

Not every admirer of a crescent moon will unerringly say whether it is new or, on the contrary, on the wane. The difference between the two is that the crescents bulge in opposite directions. In the Northern Hemisphere the new crescent moon always has its convex side to the right, and the old moon to the left. How, then, are we to remember for sure which side either crescent faces?

Join the horns of the crescent by an imaginary straight line. The result will either be the Latin letters "d" or "p". The "d" is the first letter in the word "dernier" (last), thus indicating the last quarter, or the old moon. The "p" is the first letter in "premier" (first) which shows the Moon in its first quarter, a new moon (Figure 30). The Germans also have a rule linking the shape of the moon with definite letters.

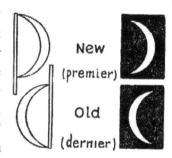

Fig. 30. A simple way to distinguish between the new (growing) moon and the old moon.

But these rules hold only in the Northern Hemisphere, since in Australia or the Transvaal, the reverse is the case. But even in the Northern Hemisphere they could be unsuitable, say, in southern latitudes. In the Crimea and Transcaucasia the crescent and half moon are strongly tilted, lying altogether on the side still further south. Near the equator the crescent is seen above the horizon either like a gondola floating on the waves (the "moonboat" of Arabian tales) or as a bright arch. Here the French device will not do since the letters "p" and "d" cannot be formed from the reclining crescent. Small wonder the ancient Romans dubbed the tilted moon "fallacious" (Luna fallax). To avoid mistaking the moon's phases then, we should turn to astronomical signs: the new moon is seen after dark in the western part of the sky, while the old is seen towards morning in the east.

The Moon on Flags

Question

Figure 31 shows the old Turkish flag with crescent moon and star. This prompts the following questions:

Fig. 31. The old Turkish flag.

(1) Which crescent moon does the flag show, new or old?

(2) Can one see crescent and star in the heavens in the position given on the flag?

Answer

(1) Recalling the devices suggested above and that the flag belongs to a country situated in the Northern Hemisphere, we can conclude that the Moon on the flag is the old one.

(2) The star cannot be seen inside the Moon's full disc (Figure 32a). All heavenly bodies are, much farther away than the Moon and hence can be eclipsed by it. They can be seen only outside the rim of its darkened part, as depicted on Figure 32b.

Curiously enough, the present Turkish flag, which also depicts crescent and star, sets the latter away from the crescent exactly as on Figure 32b.

Fig. 32, a and b. Why stars cannot be seen between the Moon's horns

Fig. 33. There is an astronomical error in this picture. What is it? (See text for answer.)

The Riddle of the Lunar Phases

The Moon receives its light from the, Sun and therefore the convex of the lunar crescent should, naturally, face the Sun. Artists, however, often forget this. At exhibitions one sometimes sees landscapes depicting a half moon with its right face towards the Sun. Sometimes one even sees a crescent Moon with horns facing the Sun. (Figure 33).

We should say, incidentally, that to draw the new Moon correctly is not as simple as it seems. Even experienced masters of the brush paint the outer and inner arcs of the crescent in the form of semicircles (Figure 34b). Yet, only the outer arc is a semicircle, the inner one is half ellipse, or a semi-circle (the terminator) seen in the perspective (Figure 34a).

Nor is it easy to set the crescent moon in its right position in

Chapter 2

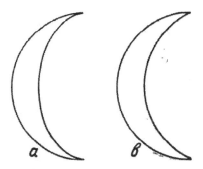

Fig. 34. How should one draw (*a*) and
not draw (*b*) a crescent moon.

the heavens. The half moon and the crescent often take rather perplexing positions with respect to the Sun. It would seem that since the Moon is illuminated by the Sun, a straight line joining the Moon's cusps should be at right angles to the Sun's bisecting ray (Figure 35). In other words, the centre of the Sun should be at the end of a perpendicular drawn from the centre of the straight line joining the Moon's horns. This rule, however, holds only for the narrow crescent moon. Figure 36 shows the Moon's position at different phases with respect to sun-rays. The impression is that the rays seem to bend before they reach the Moon.

Fig. 35. The crescent moon with respect to the Sun.

The answer to the riddle is this: the day coming from the Sun towards the Moon is actually perpendicular to the line joining the Moon's horns and draws a straight line in space. However, our eye sees in the sky not this straight

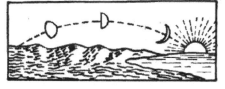

Fig. 36. At different phases the Moon is seen in these positions with respect to the Sun.

line but its projection on the caved-in bowl of the heavens, viz., a curved line

That is why the Moon seems to be "hanging the wrong way" in the sky. The artist should study these peculiarities and know how to depict them.

The Double Planet

The Earth and the Moon form the double planet. They are entitled to this name as our satellite is prominent among other counterparts for its proportionately large size and mass compared with its primary. The solar system actually has larger and heavier satellites. But compared to their primaries they are relatively much smaller than the Moon with respect to the Earth. Indeed, the Moon's diameter is more than a quarter of the Earth's. The diameter of the biggest satellite, Neptune's Triton, is but a tenth of its primary's diameter. Furthermore, the Moon's mass is 1/81 of the Earth's mass, while the solar system's heaviest satellite, Jupiter's III, is less than 1/10,000 of the mass of its primary.

The table below shows the fraction of the masses of the primaries possessed by the masses of the biggest satellites.

Primary	Satellite	Mass(fraction of primary mass)
Earth	Moon	0.0123
Jupiter	Ganymede	0.00008
Saturn	Titan	0.00021
Uranus	Titania	0.00003
Neptune	Triton	0.00129

The Moon leads the table for size of the fraction of the primary's mass.

The third circumstance entitling the Earth-Moon system to the name of "double planet" is their close proximity. Many planetary satellites rotate at distances much further away; some Jovian satellites (i.e., IX, Figure 37) are 65 times farther away.

Fig. 37. The Earth-Moon system compared with the Jovian family. (Scale has been disregarded in depicting the actual dimensions of the celestial objects.)

Here we encounter the interesting fact that the Moon's path around the Sun differs but slightly from that of the Earth. This seems incredible when one considers that the Moon is nearly 400,000 km away from the Earth. But we should not forget that in one revolution of the Moon around the Earth, the Earth itself covers, together with the Moon, roughly 1/13th of its yearly journey, viz., 70,000,000 km. Suppose the Moon's circular orbit of 2,500,000 km elongated thirtyfold. What would remain of its round form? Nothing at all. That is why the Moon's path around the Sun almost merges with the Earth's orbit, diverging only in 13 hardly noticeable bulges. By a simple calculation that we shall not bother you with, we, could prove that the Moon's path is always concave towards the Sun, roughly resembling a figure with 13 sides and 13 rounded angles.

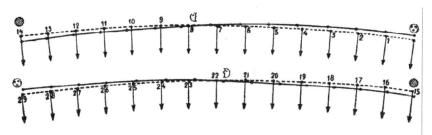

Fig. 38. The monthly passages of the Moon (solid line) and the Earth (broken line) about the Sun.

Figure 38 gives you an exact picture of the paths of the Earth and the Moon for one month. The broken line shows the Earth's path, the solid line that of the Moon. Being so close, a drawing of a very large scale was needed to show them apart. On this drawing the diameter of the Earth's orbit is half a meter. If, however, it had been shown a 10 cm, the greatest divergence between the two paths would have been less than the thickness of the lines depicting them.

An examination of this drawing shows that the Earth and the Moon revolve around the Sun in nearly one and the same orbit, and that astronomers have quite justly named this system the "double planet."[1]

Thus a solar observer would see the Moon's path as a slightly curved line, almost dovetailing with the Earth's orbit. This in no way contradicts the slightly elliptical path of the Moon's motion with respect to the Earth.

The reason is, of course, that in observing from the Earth we do not see the Moon shift together with the Earth along the Earth's orbit, as we ourselves participate, in this motion.

Why Doesn't the Moon Fall onto the Sun?

This looks like a naive question. Why should the Moon fall onto the Sun? The Earth, seemingly, exerts greater attraction than the distant Sun, and, therefore, compels the Moon to revolve around itself.

The reader who thinks in this way will be astounded to learn that actually the reverse is the case; the Sun, not the Earth, exerts greater attraction for the Moon!

Calculations will prove the point. Let us compare the forces

-1- Close scrutiny of the drawing shows that the Moon's motion is not depicted as strictly even. This is really so. The Moon revolves around the Earth along an elliptical path, having the Earth as its focus. Therefore, according to Kepler's second law, it moves faster when closer to the Earth than when farther away. The eccentricity of the lunar orbit -being 0.055- is rather great.

attracting the Moon -Sun and Earth. Both forces depend on two factors, the size of the attracting mass and its distance from the Moon. The Sun's mass is 330,000 times bigger than the Earth's; that is exactly the number of times greater the Sun's attraction would be for the Moon were the distances from the Moon the same. However, the Sun is roughly 400 times farther away than the Earth. The force of gravitation diminishes in proportion to the squared distance; hence the Sun's gravitational force should be 400^2, i.e., 160,000 times less. Consequently the Sun's gravitational force is greater than that of the Earth by 330,000/160,000, i.e., more than double.

Hence, the Sun attracts the Moon with twice the force of the Earth, Why, then, doesn't the Moon fell onto the Sun? Why does the Earth, nevertheless, compel the Moon to revolve round it, while the Sun fails to get the upper hand?

The Moon does not fall onto the Sun for the same reason the Earth does not do so; it revolves around the Sun together with the Earth and the Sun's gravitational pull is fully expended in keeping these, two bodies from a straight path, in making them take a curved orbit, i.e., in converting straight-line motion into curved-line motion. A glance at Figure 38 shows that this is so.

Perhaps some readers still have their doubts. How does all this come about? The Earth attracts the Moon to itself; but the Sun attracts the Moon with greater force; nevertheless, instead of falling onto the Sun, the Moon revolves round the Earth? This would indeed be strange if the Sun attracted the Moon alone. The point is that it attracts Moon and Earth, the whole of this "double planet," without interfering, one might say, in the domestic relations of the couple. Strictly speaking, the Sun attracts the common centre of gravity of the Earth-Moon system; it is this centre that revolves around the Sun under the influence of solar attraction. It is located at a distance of 2/3 the Earth's radius from the centre of the Earth towards the Moon. The Moon and the Earth's centre revolve around the common centre of gravity, making one complete revolution in one month.

The Moon's Visible and Invisible Faces

Among stereoscopic effects none is so startling as the appearance of the Moon. One sees that the Moon is really ball-shaped, whereas in the heavens it gives the impression of being flat like a tea-tray.

Few, however, have even the slightest idea of how difficult it is to get a stereographic photograph of our satellite. To do so one must be well acquainted with the caprices of its motions.

The Moon revolves about the Earth with one and the same face presented to it all the time. In doing so, it rotates on its axis, the two motions coinciding in time.

Figure 39 shows an ellipse depicting the Moon's orbit. The drawing deliberately exaggerates the elongated character of the lunar ellipse; actually the eccentricity of the lunar orbit is 0.055 or 1/18th. An exact reproduction of the Moon's orbit in a small drawing, which would enable the eye to distinguish it from a circle, is unthinkable, for if the major semi-axis were one, meter long, the minor semi-axis would be only

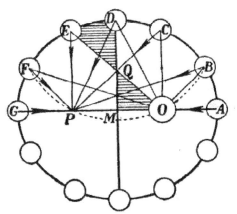

Fig. 39. How the Moon follows its orbit round the Earth (see text for details).

1.5 mm. shorter; the Earth would be but a mere 5.5 cm. from the centre. Hence, for explanatory purposes the drawing shows an elongated ellipse.

Imagine, then, the ellipse on Figure 39 as the Moon's orbit around the Earth. The Earth is located at point O, one of the foci of the ellipse. Kepler's laws apply not only to planetary motions about the Sun but also to motions of satellites around their primaries, and in particular to lunar revolutions. According to Kepler's second law, in a quarter of a month the Moon traverses the distance AE, where OABCDE is one-fourth the area of the ellipse, viz., the area MABCD (the equality of the areas OAE and MAD on our drawing is confirmed by the approximate equality of the areas MOQ and EQD). Thus, in one quarter of a month the Moon journeys from A to E, However, in contrast to revolution about the Sun, the rotation of the Moon, like planetary rotation in general, is even; in one quarter of a month it turns exactly 90°. Therefore, when it reaches E, its radius towards the Earth at point A will sweep a 90° arc and be directed not at point M but at some other point left of M, near its orbit's second focus, P. As the Moon slightly turns its face away from the terrestrial observer, he will see, on the right side, a narrow strip of its hitherto unseen hemisphere. At point F the Moon shows the earthly observer a still narrower band of its usually invisible side, as the angle OFP is less than the angle OEP. At point G, the orbit's "apogee," the Moon takes up the same position with respect to the Earth as at A, the "perigee." In its further course the Moon turns away from the Earth, this time in the opposite direction, presenting another strip of its invisible side to our planet. This strip at first widens, then narrows, and at last, at point A, the Moon resumes its old position.

We have seen that, due to the elliptical form of the Moon's course, our satellite does not strictly present one and the same face to the Earth. It invariably turns one and the same face not towards the Earth but towards the other focus of its orbit. To us it seems to sway around its central position like a pair of scales; hence the astronomical term, "libration," from the Latin word "libra" meaning "balance." The extent of libration at each point is measured by the corresponding angle, for example, at point E, libration is equal to angle OEP. The greatest libration is 7°53' or almost 8°.

It is interesting to follow the growth end diminution of the angle of libration, as the Moon moves along its orbit. Prick point

D with the needle-leg of a pair of compasses and describe an arc, passing through the foci O and P. This arc will intersect the orbit at points B and F. The angles OBP and OFP, being inscribed angles, are equivalent to half the central angle ODP. Hence we find that in the Moon's passage from A to D, libration speedily reaches half its maximum at point B and then gradually increases; between D and F libration diminishes slowly at first and then more rapidly. In the second half of the ellipse the libration varies similarly, but contrariwise. (The degree of libration at each point of the orbit is roughly proportional to the Moon's. distance away from the ellipse's major axis.)

This lunar swaying is called longitudinal libration. Our satellite is subject to yet another kind of libration, known as latitudinal libration. The plane of the lunar orbit is inclined 6½° towards the plane of the lunar equator. That is why in one case we see the Moon slightly from the south, and in the other, slightly from the north, getting a peep, over its poles, at its "invisible" hemisphere. This latitudinal libration reaches as much as 6½°.

I shall now try to explain how the astronomer avails himself of the slight swaying of the Moon about its centre to obtain a stereoscopic photograph. The reader has probably already guessed that to do this one must catch two positions of the Moon having a big enough angle between them.[2] At points A and B, B and C, C and D, etc., the Moon's position with respect to the Earth varies to the degree where stereoscopic photographs are possible. Here, however, we are faced with another difficulty; in these positions the difference in the age of the Moon some 36 to 48 hours, is so great that its surface strip near the terminator on the first photograph has already emerged from darkness. This will not do for stereoscopic pictures since the strip will have a silver shine. This confronts us with the knotty task of catching identical Moon phases differing in degree of longitudinal libration so that the illuminated circle cover one and the same lunar surface. But even this is not enough; the two positions must have, in addition, an identical latitudinal libration.

-2- A 1° angular turn of the Moon suffices to obtain a stereoscopic photograph (For greater detail, see my *Physics for Entertainment*.)

You will now be aware how hard it is to obtain good stereo-photographs of the Moon. And you will not be surprised to find that it often takes years to make the second picture of the stereoscopic pair.

But the reader will hardly indulge in lunar stereo-photography. Our explanation is designed not for practical purposes but to point to the specific features of lunar motion which enable astronomers to see *a* small strip on the side of our satellite usually inaccessible to observation. The lunar librations enable us to see not half, but 59% of the entire lunar surface. The remainder is absolutely beyond our vision, and no one knows what it looks like; we can but presume that it does not differ in essentials from the visible port. Clever attempts have been made, by continuing some of the lunar ranges and the bright belts, to trace, tentatively, some of the details of the inaccessible hemisphere. But so far we are in the realm of guesswork. We say "so far" because we have well advanced towards flying around the Moon on a machine capable of overcoming the Earth's gravitation and journeying into outer space. The day when this daring enterprise will be accomplished is not so very far away. One thing we do know: the frequently voiced thesis of the Moon's invisible hemisphere having an atmosphere is absolutely untenable; it is in crying contradiction to the laws of physics. If one side of the Moon has no atmosphere, how can the other have any? (We shall revert to this question later.)

A Second Moon and the Moon's Moon

Newspapers have reported from time to time that one or another observer has established the existence of a second satellite of the Earth, a second Moon. Although these claims have never been confirmed, the subject is not without interest.

The question of the Earth having another satellite is not a new one -it has a long history. Those who have read Jules Verne's

De la terre à la lune will recall that the author mentions a second Moon which was so small and fast, that it could not be seen from the Earth. The astronomer Petit, says Jules Verne, suspected its existence and defined its revolution around the Earth as being 3 hrs 20 min. It was supposed to be 8,140 km. distant from the Earth. Curiously enough, in an article discussing the astronomy in Jules Verne's books, the English magazine *Science* expressed the view that the reference to Petit and Petit himself were pure invention. Certainly no encyclopedia has ever mentioned an astronomer of this name. Nevertheless, the novelist was not guilty of invention. In the fifties of the last century Toulouse Observatory director Petit actually mentioned the existence of a second moon, a meteorite having a 3 hrs 20 min. period of revolution, and located not 8,000 but 5,000 km. from the Earth. Only a few astronomers shared this view at the time. Subsequently it was consigned to oblivion.

Theoretically, there is nothing unscientific in presuming that the Earth has another, tiny satellite. But any celestial body of this order should be seen not only on the rare occasions of its seeming transits across the disc of the Moon or the Sun. Even if it rotated so close to the Earth that at each revolution it plunged into the Earth's shadow, it would still be seen in the morning and evening as a bright star shining in the Sun's rays. Because of its rapid motion and frequent recurrences, this star would certainly have attracted the attention of a host of observers. Nor would a second moon elude the astronomer's eye during a total solar eclipse.

In short, if the Earth did have another satellite it would be seen fairly often. But it is incontestable that no such sight has ever been observed.

Another suggestion is that the Moon may have its own little satellite, the Moon's moon. It would, however, be extremely difficult to vouch directly for the existence of a lunar satellite of this kind. Here is what astronomer Moulton has to say on the, subject:

"In full Moonshine, the light of the Moon or the Sun prevents detection of a tiny body in its vicinity. Only in lunar eclipses would the Moon's satellite be lit by the Sun and the neighboring heavens be free of the Moon's dispersed light. Thus, only in lunar eclipses could we expect to discover a small body circulating around the Moon. Investigations have been conducted to this end but have so far yielded no tangible results."

Why Is There No Air on the Moon?

This question falls into the category of those which become clear when, so to speak, we turn them upside down.

Before trying to find out why the Moon has no air, let us see why our own planet is enveloped by an atmosphere. Remember that air, like any gas, is a chaos of molecules dashing about in all directions. Their average velocity at 0°C is about 0.5 km/sec, that of a rifle bullet. Why, then, do they not escape into space? For the same reason that the bullet does not escape into space. Having expended the energy of their motion in overcoming gravitation, the molecules again fall out onto the, Earth. Imagine in the vicinity of the Earth's surface a molecule flying straight upwards at 0.5 km/sec. What altitude would it reach? This is easy to compute, as velocity v, altitude h, and acceleration of gravity g are interlinked in the formula:

$$v^2 = 2\,gh$$

Let us replace v by its value, 500 m/sec and g by 10 m/sec²; the result will be:

$$250{,}000 = 20h, \text{ whence}$$

$$h = 12{,}500 \text{ m. or } 12.5 \text{ km.}$$

But, if air molecules cannot get higher than 12.5 km., the question arises: whence the air molecules above this boundary? The oxygen of our atmosphere is formed near the Earth's surface (from carbon dioxide through plants). What force, then, raises and keeps them at a height of 500 and more kilometers, where traces of air have been definitely established. Physics supplies the identical answer that we would get from any statistician were we to ask: "If the average expectation of human life is 40 years, where do 80-year-olds come from?" The point is that our calculation is good for the mean molecule, not the real one. The mean molecule has a per second velocity of half a kilometer, but as for real molecules, some move slower, and some faster than average velocity. True, the proportion of molecules whose velocities appreciably diverge from the mean is not great and rapidly declines with the increase in the divergence. Of the total number of molecules in a definite volume of oxygen at 0°C only 20% possess a velocity of between 400 to 500 m/sec., roughly the same number have a speed ranging from 300 to 400 m/sec., 17% between 200 and 300 m/sec., 9% between 600 and 700 m/sec., 8% between 700 and 800 m/sec., and 1% between 1,300 and 1,400 m/sec. A very minute portion (less than 1/1,000,000[th]) of the molecules has a velocity of 3,500 m/sec., enough for them to soar as high as 600 km. For, $3500^2 = 20h$, from whence h= 12,250,000:20, i.e., over 600 km.

Now we know why there are oxygen particles at an altitude of hundreds of kilometers above the surface of the globe; this stems from the physical properties of gases. However, molecules of oxygen, nitrogen, vapor and carbon dioxide do not have a velocity strong enough to escape from the Earth altogether. This would necessitate a velocity of no less than 11 km/sec. Only isolated molecules of the above-mentioned gases possess these velocities at low temperatures. That is why the Earth holds its blanket of air so tightly. It has been computed that it would take years of an order of 25 figures for even half the store of hydrogen, the lightest gas in the terrestrial atmosphere, to escape. Millions of years would not make any change whatever in the composition and mass of air around us.

After this, little need be said in order to explain why the Moon cannot retain a similar atmosphere. The Moon's gravity is 1/6 of the Earth's. Accordingly, the velocity needed to overcome its gravitation is only 2,360 m/sec. As the velocity of oxygen and nitrogen molecules at moderate temperatures may be more, it follows that the Moon would continually lose its atmosphere, were it to form there. With the escape of the fastest molecules, other molecules would acquire the critical velocity (a consequence of the law of distribution of velocity among gas particles), and increasing numbers of particles of the enveloping air would fall out irrevocably into space. With the lapse of a sufficiently long interval, negligible in terms of the universe, all the air would escape from the surface of a celestial body with such weak attraction.

It can be proved mathematically that if the average velocity of the molecules in a planet's atmosphere were, but a third of the limit (i.e., 2,360:3=790 m/sec. for the Moon), an atmosphere of this kind would dwindle to half in the space of a few weeks. (An atmosphere will firmly cling to a celestial body only if the mean velocity of its molecules is a fifth of the peak.)

The idea, or rather the dream, has been suggested that one day when man reaches the Moon he will surround it with an artificial atmosphere and make it fit for human habitation. But after what has been said, readers will realize that an enterprise of this nature is out of the question. The fact that our satellite lacks an atmosphere is not fortuitous, not a caprice of nature, it is the logical result of physical laws.

We can see that the reasons explaining the absence of atmosphere on the Moon also explain its absence on all celestial bodies with a weak gravitational force, for example, the asteroids and most planetary satellites.[3]

-3- In 1948 the Moscow astronomer Y. N. Lipsky apparently found traces of an .atmosphere on the Moon. But the general mass of the Moon's atmosphere cannot the more than 1/100,000[th] that of the Earth.

Lunar Dimensions

Figures, of course, tell us a great deal about these, say, about the Moon's diameter (3,500 kilometers), its surface and volume. But figures, while indispensable for calculations, cannot transmit a picture of the dimensions that we should like to see with the mind's eye. Here concrete parallels would be much better.

Let us compare the lunar continent -the Moon is all one continent- with the continents of the Earth (Figure 40). This will tell us more than the abstract notion of the Moon's entire, surface being a fourteenth of the Earth's. In square kilometers the surface of our satellite is but slightly less than the area of the two Americas, while the lunar face presented to the Earth and accessible to observation is almost the exact area of South America.

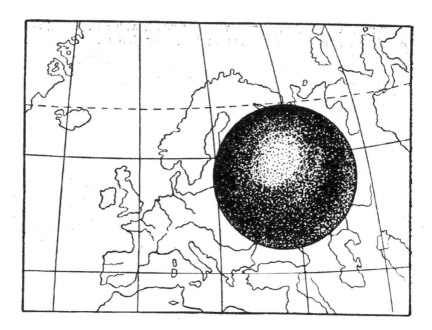

Fig. 40. The Moon compared with the continent of Europe. (Beware of thinking, however, that its surface is less than that of Europe.)

To illustrate the dimensions of the lunar "seas" compared with terrestrial bodies of water, Figure 41 has superimposed upon a map of the Moon the outlines of the Black and Caspian seas, on the same scale. We can see at a glance that the lunar "seas" are not very large, though they account for a sizable portion of the Moon's disc. The lunar Mare Serenitatis, for instance, is 170,000 km² in area -roughly 2/5 of the Caspian.

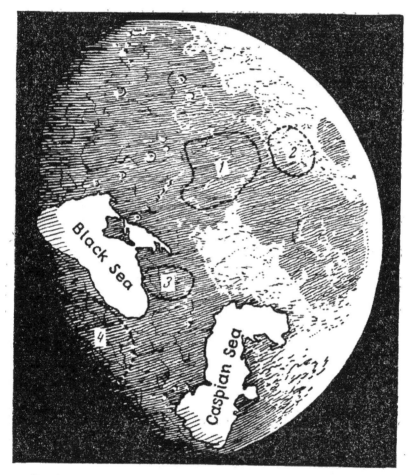

Fig. 41. Terrestrial seas compared with lunar seas. If the Black and Caspian seas were transferred to the Moon they would take precedence over all the lunar maria. (*1*—Mare Nubium, *2*—Mare Numorum, *3*—Mare Vaporum, *4*—Mare Serenitatis.)

On the other hand the lunar craters are vaster by far than anything the Earth can show in this respect. The Grimaldi Crater, for instance, covers an area larger than Lake Baikal and could encompass a small country, stay, Belgium or Switzerland.

Lunar Landscapes

So many photographs of the Moon's surface have been repro-
duced in books that I suppose all my readers to have an idea of
the typical features of lunar relief, such as the craters (Figure 42),
the "cirques." Some, probably, will have seen lunar mountains
through a small telescope -an eyeglass with a 3 cm lens is enough.

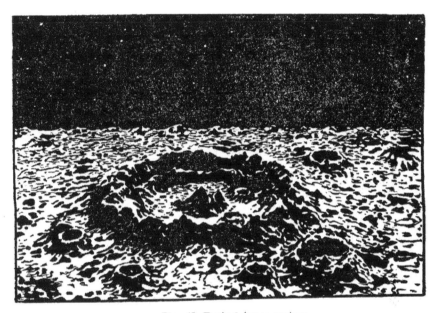

Fig. 42. Typical lunar craters.

However, neither photographs nor telescopic observations
furnish any idea of what the lunar surface would look like to
an observer on the Moon itself. From his vantage point near
the lunar mountains the observer would see them from quite a
different angle. It is one thing to look at an object from a high
elevation, and, quite another, from nearby. A few examples will
illustrate the point. To the terrestrial observer the Eratosthenes
Mountain has the appearance of a "cirque" with the peak inside.
Seen through the telescope it is precipitous and thrown into relief by the clearly defined, marked shadows. Glance, however,
at its profile (Figure 43). You will see that, compared with the
vast 60 km diameter of the cirque, the wall and inner cone are
very low, with the slopes attenuating their height. Now imagine
yourself roaming around inside this crater. Don't forget that in

diameter it is equivalent to the distance between Lake Ladoga and the Gulf of Finland. You would scarcely notice the circular form of the walls, while the concavity of the ground would obscure its base, as the lunar horizon is twice narrower than the Earth's (since the Moon's diameter is one-fourth of the Earth's). On the Earth a man of average height, standing in the centre of a level plain has a vision of 5 km. This derives from the formula for the distance of horizon[4] where D is the distance in kilometers, h eye's height in kilometers and R planet's radius in kilometers.

Fig. 43. A large crater in profile.

Replacing the letters in this formula with the respective figures for the Earth and the Moon we find that for a man of average height the horizon distance is 4.8 km. on the Earth and 2.5 km on the Moon.

Figure 44 shows what an observer would see inside the Moon's Metropolitan crater. (The landscape shown is for another of its major craters -the Archimedes.) Does not this vast plain with a chain of hills away to the horizon differ greatly from one's mental picture of a "lunar crater?"

After climbing the wall to the other side of the crater the observer would again see not what he had anticipated. The outer slope of the crater (see Figure 43) is so "unprecipitous" that the observer would not take it for a mountain at all. And the chief thing is that he would never take the range to be circular and with a round crater. To get the needed picture he would have to cross the crest, and even then, as I have already said, he would

-4- For computation of the distance of horizon, see my Geometry for Entertainment -the chapter headed "Where Heaven and Earth Meet"

not be rewarded with anything remarkable.

Fig. 44. What anyone would see from the centre of a large lunar crater.

In addition to its immense craters the Moon has a multitude of smaller cirques easily discernible even when the observer is nearby. Their height, however, is negligible, and again he would hardly see anything out of the ordinary. On the other hand the lunar ranges, named like those on the Earth -the Alps, the Caucasus, the Apennines, and so on, vie with their terrestrial counterparts, being some 7 and 8 km high. On the comparatively small Moon they are a rather impressive sight.

The Moon's lack of an atmosphere and the related sharp contrast between light and shadow produce, in telescopic observation, an interesting illusion:

Fig. 45. Side lighting gives a split pea a long shadow.

the slightest unevenness is accentuated and thrown info bold relief. Place a split pea round side up on a piece of paper. It seems small, doesn't it? But look at the long shadow it casts (Figure 45). Due to side-lighting a shadow on the Moon may be 20 times longer than the height of the object casting it. This is a boon to astronomers because the long shadows have facilitated telescopic observation of lunar objects as low as 30 m. in height. But the same factor makes us exaggerate the uneven nature of lunar terrain. For instance Mt. Pico, seen through the telescope, stands out in such sharp relief that one involuntarily takes it to be a steep, jagged peak (Figure 46). And as a matter of fact that was how it had been depicted in the past. But, observing it from the Moon's surface we would get quite another picture, the one shown on Figure 47.

Fig. 46. In the telescope Mt. Pico has the appearance of a steep, pointed cliff.

Fig. 47. The Moonman sees Mt. Pico as a slope.

On the other hand, we underrate other features of the Moon's surface. The telescope shows us slight, barely noticeable crevices

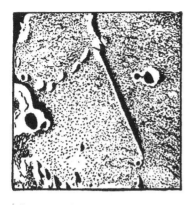

Fig. 48. The Moon's so-called "Straight Wall" (as seen through the telescope).

which, seemingly, are minor features of the lunar landscape. But by transporting ourselves to the surface of our satellite we would see at our feet a deep, dark abyss stretching far to the horizon. Here is another example. The Moon has what is known as the "Straight Wall," a steep shelf intersecting one of its plains. When we see this wall on a chart (Figure 48), we forget it is 300 m high; but were we at its base we would be overwhelmed by its grandeur. On Figure 49 the artist has endeavored to depict this wall as seen from below; its end disappears into the horizon, for it is about 100 km long! And in exactly the same way the barely discernible clefts seen on the Moon's surface through a powerful telescope must be, in the nature of things, huge abysses (Figure 50).

Fig. 49. This is how anyone standing at the foot of the "Straight Wall" would see it.

Fig. 50. At the edge of a lunar "crack."

Lunar Heavens

Canopy of Black

If a man from the Earth were to find himself on the Moon, his attention would be drawn right away by three unusual things.

The strange color of the lunar heavens by day would strike the eye immediately, as instead of the usual blue the canopy overhead would be absolutely black, despite the bright sunshine. It would show myriads of stars, all plainly seen, but not one of them twinkling. The reason: Absence of atmosphere.

"The blue canopy of a clear and pure sky," says Flammarion in his picturesque language, "the tender flush of dawn and the majestic flame of evening sunset, the entrancing beauty of deserts, the mist-shrouded far-stretching fields and meadows, and you, the mirror-like lakes which reflect the distant azure heavens, with waters as deep as infinity -your very existence and all your beauty depend solely on the flimsy casing enveloping the Earthly globe. Without it, not one of these pictures, not one of these luxuriant colors would exist. Instead of azure blue heavens you would be surrounded by the black of an endless void; instead of majestic risings and settings of the sun, the days would abruptly change with no gradual transition to nights, and vice versa. Instead of the tender half-light which prevails wherever the blinding rays of the sun do not fall directly, there would be light only in those places directly illuminated by our diurnal luminary; deep shadow would prevail in all other places."

The atmosphere need be rarified only slightly for the blue of the sky to become darker. The crew of the Soviet "Osoavi-akhim" stratosphere balloon which met such a tragic fate saw at an altitude of 21 km an almost black sky. The description of nature's lighting in the excerpt above is completely true of the Moon, with its black heavens, absence of morning and evening twilights, glaringly illuminated places, and places wrapped in darkness, devoid of any half-tones.

The Earth as Seen in Lunar Skies

The second remarkable lunar spectacle would be the Earth's huge disc suspended in the sky. We would find it queer for the Earth, which when we took off for our ascent to the Moon seemed to be *beneath* us, suddenly to show itself *above* our heads.

The universe, naturally has no common below and above for all the worlds. So there is no reason to be surprised when we find the Earth that we left below appears overhead when we reach the Moon.

The Earth's disc hanging in the lunar heavens is of enormous dimensions; its diameter is roughly four times larger than the familiar disc of the Moon seen from the Earth. This is the third amazing fact that awaits the lunar traveler. If on moonlit nights our landscape is rather well illuminated, then night on the Moon in the full rays of light from the Earth with its disc 14 times larger than the Moon, should be exceedingly bright. The brightness of a luminary depends not only upon its diameter but also on the reflecting capacity of its surface. In this respect the Earth's surface is six times better off than that of the Moon;[5] hence, full "Earthshine" should light up the Moon at least 90 times more powerfully than the full Moon lights up the Earth. In full "Earthshine" the man in the Moon could read even small print. The Earth illuminates the Moon's surface so brightly that we, 400,000 km. away, can see the Moon's night bound parts as a hazy twinkling inside a narrow crescent; this is known as the Moon's "ash-light." Imagine 90 full Moons shilling down and add to this our satellite's lack of an atmosphere to engulf part of the light, and you will get an idea of the enchanting, fairylike spectacle of nocturnal lunar landscapes basking in full "Earthshine."

Would the lunar observer distinguish on the Earth's disc the outlines of continents and oceans? There is a widely held mistaken view that to the man in the Moon the Earth resembles a school globe. That, at any rate, is how it is drawn by artists when they want to show it in space, covered with the contours of continents, polar ice-caps and other details. This is pure imagination. The observer would not see anything of the sort. To say nothing of the clouds which usually obscure half the Earth's surface, our atmosphere itself vigorously disperses the sunrays, hence the Earth should seem just as bright and impenetrable to observation as Venus. G. A. Tikhov, a Pulkovo astronomer, who investigated this matter, says: "Looking at the Earth from space we would see a disc colored like a rather whitish sky and would hardly detect any details of the actual surface. Most of the sunshine falling on Earth is dispersed in the atmosphere with all its admixtures before it reaches the Earth's surface. And that part of it reflected from the surface is also greatly weakened by further atmospheric dispersion."

Thus, while the Moon distinctly reveals its surface details, the

-5- The ground on the Moon, therefore, is not white as commonly believed; it would be more correct to say that it is dark. This does not contradict the fact that it shines with a white light. "Sunshine, even when reflected back from a black object, remains white. Even if the Moon were draped in the blackest of velvet, it would still be seen as a silvery disc," says Tyndall in his book on light. The capacity of the lunar ground to disperse the solar rays is on the average the equivalent of the dispersion capacity of dark volcanic rock.

Earth, on the contrary, hides its face from the Moon, and the entire universe for that matter, under a shining veil of atmosphere.

That, however, is not the only difference between the Moon and the Earth. In our sky the Moon rises and sets, travelling together with the canopy of stars. But the Earth does not follow suit in lunar skies; It neither rises nor sets there, and does not take any part whatever in the exceedingly slow parade of the stars. It hangs almost immobile in the sky, in a fixed position for every part of the Moon, while the stars slowly glide on behind it. This is due to the peculiar feature of lunar motion already examined, the gist of which is that the Moon always presents one and the same face to the Earth. To the lunar observer the Earth would seem to be almost fixed in the sky. For instance, should the Earth be at zenith of a lunar crater it will never leave it. If seen on the horizon it will remain there perpetually. Only the lunar librations mentioned above somewhat disturb this immobility. The starry heavens accomplish, behind the Earth's disc, a slow revolution of 27.33 days, the Sun rides the sky in 29 ½ days, the planets make similar motions, and only the Earth hangs almost immobile in al black sky.

But, while fixed in one place, the Earth quickly makes 24-hour axial rotations, and if our atmosphere were transparent the Earth would be a most convenient timepiece for the passengers of future space-ships. Furthermore the Earth has the same phases as those shown by the Moon in our sky. Hence, the Earth does not always shine in lunar skies as a full disc: sometimes it appears as a half-circle, sometimes as a wider or narrower crescent and sometimes as an incomplete circle, depending on the part of its sunlit hemisphere presented to the Moon. By charting the related positions of the Sun, the Earth and the Moon, you will easily perceive that the Earth and the Moon present opposite phases to one another.

When we see the new Moon, the man in the Moon sees the Earth's full disc, a "full Earth" and, conversely, when we see the full Moon, the latter sees a "new Earth" (Figure 51). When we see the narrow crescent of the new Moon, the man in the Moon sees the Earth on the wane, with exactly the same crescent, as the Moon presents to us at the moment, missing. Incidentally, the Earth's phases do not interchange so abruptly as those of the Moon; the Earth's atmosphere erases the terminator and creates that gradual transition from day to night and vice versa, which we on Earth see as twilight.

Fig. 51. A "new Earth" on the Moon. The Earth's black disc is hemmed by the bright rim of its shining atmosphere.

There is one more difference between the phases of the Earth and the Moon. On Earth we never see the Moon at new Moon proper. Although in this case it is usually above or below the Sun (sometimes as much as 5°, i.e., 10 times its diameter), in a position allowing the Moon's sunlit limb to be seen, it is inaccessible to vision because the Sun's brilliance absorbs the modest shine of the silvery thread of the new Moon. As a rule we see the new Moon when it is around two days old, that is, when it is far enough away from the Sun. Only in rare cases, in spring, do we see it when it is a day old. This would not be the case if we were observing the "new Earth" from the Moon, where there is no air to disperse the Sun's rays and create the customary halo around our diurnal luminary. There, the stars and planets are not lost in the Sun's rays, they stand out plainly in the sky, in the Sun's immediate vicinity. Thus, when the Earth is not directly in front of the sun (i.e., not in eclipse, but somewhere above or below), it will always be seen against our satellite's black, star-spangled sky in the form of a narrow crescent with the cusps turned away from the Sun (Figure 52). According as the Earth moves leftwards from the Sun, the crescent seems to follow suit.

Fig. 52. The "new" Earth in lunar skies. The white disc below the crescent Earth is the Sun.

A phenomenon similar to that just described will be seen when observing the Moon through a small telescope; at full Moon we do not see our nocturnal luminary as a full disc; as the centers of Moon and Sun are not on a straight line with the observer's eye, the Moon's disc lacks the narrow crescent which, as a dark strip, skims leftwards along the edge of the illuminated disc according as the Moon moves right. However, the Earth and the Moon always present each other with opposite phases; at this moment the lunar observer would see the slender crescent of the "new Earth."

Fig. 53. The Earth's slow motions near the lunar skyline. The broken lines are the paths of the centre of the Earth's disc.

We have noted in passing that an effect of the Moon's librations is that the Earth is not absolutely stationary in the Moon's sky; it sways 14° in the vicinity of the mean position in the north-south direction, and 16° in a west-east direction. Wherever on the Moon our Earth is seen above the horizon, it should seem to set and then again rise, describing queer curves (Figure 53). This peculiar rising or setting of the Earth it a definite point on the horizon, without travelling across the heavens, may persist, for many Earth days.

Lunar Eclipses

I shall now add to our sketch of the lunar heavens a description of the celestial phenomenon known as an *eclipse*. There are two kinds of eclipses on the Moon, solar and "terrestrial." Although unlike the familiar solar eclipse, the first is striking in its own way. It takes place on the Moon when we see the lunar eclipse on Earth; the Earth is then set on the straight line joining the centers of the Sun and the Moon. When this happens, our satellite is submerged in the shadow cast by the Earth. Whoever has watched the Moon at this particular moment will remember that it is not totally devoid of light, nor does it vanish from vision, it is usually seen in the cherry-red rays which penetrate into the cone of the Earth's shadow. Were we to travel to the Moon at this moment and look earthwards we would easily understand the reason for this red light; in the Moon's skies; the Earth, set as it is in front of a bright but much smaller Sun, would appear as la black disc fringed with a crimson atmosphere. It is this fringe that sheds the reddish light on the shadow-darkened Moon (Figure 54).

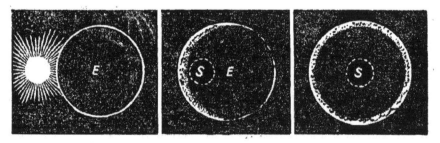

Fig. 54. A solar eclipse as seen from the Moon. The Sun *S* gradually creeps onto the Earth's disc *E*, rigidly suspended in the Moon's sky.

On the Moon solar eclipses last not the few minutes they do on the Earth, but four hours and more, or just as long as our lunar eclipses, because in substance these are lunar eclipses, only they are observed not from the Earth but from the Moon.

As for "Earthly" eclipses, they are so paltry that they hardly merit the name of eclipse. They occur when the solar eclipse is seen on Earth. The man in the Moon would then see against the huge disc of the Earth a tiny black circle, flitting across the lucky places on the Earth where the solar eclipse can be observed.

We should note in passing that eclipses like those of the Sun cannot be observed anywhere else in the planetary system. For this exclusive spectacle we are indebted to a chance factor. The Moon which comes: between us and the Sun is nearer to us than the Sun exactly the number of times that the Moon's diameter is less than that of the Sun, a coincidence unparalleled for any other planet.

Why Do Astronomers Observe Eclipses?

Due to the chance factor just mentioned, the long cone of the shadow which always accompanies our satellite touches the surface of the Earth (Figure 55). Actually the *mean* length of the cone of the lunar shadow is less than the *mean* distance between the Moon and the Earth, and were we dealing solely in mean values we would never have a total solar eclipse. They occur

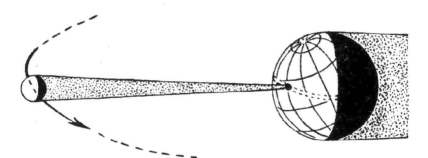

Fig. 55. The tail-end of the Moon's umbra traces the shadow-path across the surface of the Earth. It is here that solar eclipses are observed.

only because the Moon revolves about the Earth along an elliptical path which at some points is 42,200 km nearer to the Earth than elsewhere. The Moon may be anything from 356,900 to 399,100 km distant.

Flitting across the surface of the Earth, the tail-end of the Moon's shadow pencils the "strip of the visible solar eclipse." This is never wider than 300 km so that the places lucky enough to see a solar eclipse are somewhat limited. Add the duration of the total solar eclipse (a matter of minutes and never more than eight) and you will realize that a total solar eclipse is an exceedingly rare phenomenon, occurring only once every 200-300 years for any point on the globe.

For this reason astronomers literally ferret out solar eclipses, sending special expeditions to the places where it can be observed, no matter how distant they may be. The 1936 solar eclipse (on June 19) could be observed as a total eclipse only within the boundaries of the Soviet Union. For a two-minute observation, 70 scientists from 10 countries thought it worthwhile to visit our country; the labors of four of these expeditions were lost due to cloudy weather. The scale of the observations carried out by Soviet astronomers was exceedingly great -some 30 expeditions set out for the area of the total eclipse.

In 1941, despite the war, the Soviet Government sponsored a number of expeditions which stationed themselves all along the boundaries of the, total eclipse band, from Lake Ladoga to Alma-Ata. In 1947 a Soviet expedition went to Brazil to observe the total eclipse there of May 20. Observation of total solar eclipses of February 25, 1952, and June 30, 1954, assumed particularly big proportions in the U.S.S.R.

Although there are but two lunar eclipses for every three solar eclipses, they are *observed* far more often. There is a simple explanation for this apparent astronomical paradox.

A solar eclipse can be seen on our planet only in a limited

zone where the Sun is eclipsed by the Moon; within the boundaries of this narrow band the eclipse is total for some locations and partial (that is, the Sun is only partially eclipsed) for others. The timing of the solar eclipse is not the same for different places within the shadow path, not because of the difference in Urns reckoning, but because in crossing the Earth's surface the Moon's umbra covers different locations at different times.

A lunar eclipse is a different matter altogether. It can be observed simultaneously over the entire hemispheres where the Moon is visible at the given moment, that is, where it is above the horizon. The consecutive phases of the lunar eclipse set in at once all over the Earth; the only difference is in the time reckoning.

That is why the astronomer has no need to "hunt" for lunar eclipses -they come of their own accord. However, in order to "net" the solar eclipse, distant journeys are sometimes undertaken. Are these expensive expeditions for such fleeting observations worthwhile? Could we not conduct the same observations without waiting for a chance eclipse of the Sun by the Moon? Why shouldn't the astronomers artificially produce a solar eclipse by placing an opaque disc between the Sun and the telescope? We could then, without going to any great pains, observe the outlying parts of the Sun which so deeply interest astronomers during an eclipse.

However, our man-made solar eclipse would not produce the effects seen when the Moon eclipses the Sun. The point is that, before coming within range of vision the Sun's rays pass through the Earth's atmosphere where they are dispersed by particles of air. That is why we see the daytime sky as a pale blue vault, not the black star-spangled canopy we would see even in daytime were there no atmosphere. Having obscured the Sun, we remain, nevertheless, at the bottom of an ocean of air. And while we can shield the eye from the direct rays of our diurnal luminary, the atmosphere, overhead is still full of sunshine and continues to disperse its rays, thus blanketing out the stars. This would not be the case, were the screen located outside the atmosphere.

The Moon is, in a sense, a screen of this kind, a thousand times farther from the Earth than the upper boundary of the atmosphere acts as a barrier to the Sun's rays before they penetrate into the Earth's atmosphere. Consequently there is no dispersion of light in the umbra-covered path. Not completely, it is true, because some rays: dispersed by the surrounding illuminated areas, manage to penetrate into the shadow area and, for this reason, the sky during a total solar eclipse is never as dark as it is at midnight; only the brightest stars can be seen.

What do astronomers try to achieve when observing a total solar eclipse?

They first try to catch the so-called "shift" of the spectral lines in the Sun's outer layer. The lines of the solar spectrum, usually dark against the bright band of the spectrum, flash for a few seconds against a. darkened background after the Moon's disc has totally eclipsed the Sun; the spectrum of absorption becomes one of emission -the "flash spectrum." Although this phenomenon, which yields a wealth of valuable data on the nature of the Sun's outer layer, can, in specific conditions, be observed not only during an eclipse, it is so distinct during eclipses, that astronomers are eager not to let such a fortunate occasion slip.

Study of the *solar corona*, the most remarkable of all the phenomena witnessed during a total solar eclipse, is the second aim. This bright aureole, varying in size and form during eclipses (Figure 56), surrounds the absolutely black disc of the Moon with flaming protuberances or prominences. Its streamers are often several times longer than the Sun's diameter and emit a brilliance usually half that of a full Moon.

Fig. 56. As the total eclipse takes place, the "solar corona" flashes from behind the Moon's dark disc.

The 1936 eclipse produced a solar corona of exceptional brilliance, even brighter than the full Moon, which was a somewhat rare occurrence. The streamers, somewhat blurred, were three times and more the Sun's diameter; the full corona resembled la five-pointed star with the Moon's dark disc in the centre.

The nature of the solar corona has still not been completely explained. During eclipses astronomers photograph the corona, measure its brilliance and investigate its spectrum. This aids the study of its physical structure.

The third aim, advanced only in recent few decades, is to verify one of the consequences of the general theory of relativity. According to this theory, the rays of the stars passing the Sun are deflected by its powerful attraction, which should be revealed in a seeming stellar shift in the vicinity of the Sun's disc (Figure 57). The checking can be done only during a total solar eclipse.

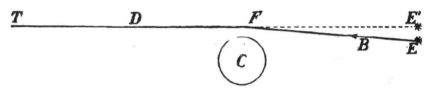

Fig. 57. A consequence of the general theory of relativity—the deflection of light by the Sun's force of gravity. According to the theory of relativity, the Earthly observer at point T sees the star E^1 along the direction of the straight line TDFE¹, whereas the star is actually situated at point E and sends its rays along the curved line $EBFDT$. If there were no Sun the ray coming from the star E towards the Earth T would travel along the straight line ET.

Measurements taken during the 1919, 1922, 1926 and 1936 eclipses did not yield, strictly speaking, any pronounced results. Experimental corroboration of the said consequence of the theory of relativity is still awaited.[6]

These, then, are the main reasons why astronomers quit their observatories and set out for remote and sometimes exceedingly inhospitable climes to observe the solar eclipse.

-6- The fact of deflection has been confirmed, but full quantitative agreement with the theory has not been established. Prof. A. A. Mikhailov's observations necessitated re-examination of some aspects of the theory.

As for the actual spectacle of the total solar eclipse, there is an excellent description of it in Korolenko's *Eclipse* -a narrative of the eclipse which took place on August 18, 1899, and which was observe in the Volga town of Yuryevets. Here are excerpts from Korolenko's story:

"The Sun sinks for a moment into a flowing curtain of haze and re-emerges from the clouds already noticeably on the wane....

"It is now visible to the naked eye, thanks to the, slight vapor still eddying in the air and attenuating the blinding glare.

"Silence. Only quick and heavy breathing is heard here and there...

"Half an hour passes. The day is almost as bright as usual; clouds obscure and reveal the Sun, now swimming, crescent-like, in the heavens.

"The young people are most excited and curious.

"Greybeards heave a sigh, old women sob hysterically, some of them even scream and groan, as if they had a toothache.

"The daylight wanes noticeably. Faces take on a frightened aspect. Human shadows lie faint and vague on the ground. A steamer going downstream slips past phantom-like. It seems, somehow, to be lighter, its colors blurred. The light, apparently, is ebbing, but without the deep shadow of evening and the play of reflected light on the lower layers of the atmosphere, this twilight seems strange and eerie. The landscape, seemingly, has dissolved into something; the grass has shed its green and the hills seem to have lost their heavy density.

"But, as long as the slender crescent rim of the Sun remains, the impression of an exceedingly dull day also remains and it seems to me that the stories about the darkness during an eclipse are, exaggerations. 'Can this still remaining minute spark of the sun, burning in the vast universe, like a forgotten last candle,' I reflected, 'really mean so much?... Can it be that when it goes out night is bound to fall suddenly?'

"But now the last spark has vanished. It flared up all of a sudden, as if forcibly bursting out of durance, scattered a shower of golden rain and was gone on the instant. Darkness swallowed the earth. I caught the fleeting moment between twilight and the fast descending darkness. It appeared in the south and, like some gigantic pall rapidly enveloped hill, river and field and, enclosing the expanses of the heavens, tucked us in and in a moment was at one with the north. From the low bank where I stood, I glanced at the crowd. The silence of the grave reigned... Human figures had merged into one dark mass....

"But this was not the ordinary night. It was so light that the eye, willy-nilly, sought the silvery moonshine penetrating the opaque blue of the ordinary night. But neither shine nor opaque blue was to be seen. It seemed as if a fine, imperceptible ash had been scattered from above upon the Earth, or if the slenderest tracery of innumerable lines hung suspended in the air. While away to the side, in the upper layers, one had the feeling of a radiant expanse of air, currents of which, flowing into our opaque darkness, merged the shadows and deprived the gloom of its shape and density. And overhead, above all the confusion of nature, clouds raced by in a wonderful panorama, an entrancing duel being fought in their midst... A round, dark, hostile body had clawed, spiderlike, into the bright sun, and the two soared upwards, far beyond the clouds. A shaft of rippling light, emerging from behind the dark shield, lent life and motion to the spectacle, while the clouds, in restless and noiseless flight, enhanced the illusion."

For the modern astronomer, the *lunar* eclipse lacks the exceptional interest of the solar eclipse. In the lunar eclipse our forefathers found convenient proof of the Earth's spherical shape. It is worthwhile recalling the role played by this notion in Magellan's voyage round the world. When, after their long and weary passage through the wastes of the Pacific, the crew, convinced that they had irrevocably parted with land, fell into despair, Magellan alone did not lose heart. "Although the Church, on the basis of the Holy Scriptures, had always affirmed that the Earth was a vast plain surrounded by water," the famous navigator's companion wrote, "Magellan reasoned thus: 'during the lunar eclipse the shadow cast by the Earth is a round shadow, hence, if the shadow is round the object which casts it must, likewise, be round.'" In old astronomical treatises we can even find drawings explaining the dependence of the shape of the lunar shadow on the shape of the Earth (Figure 58).

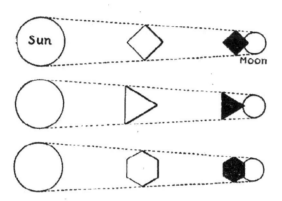

Fig. 58 An ancient drawing made to explain how the Earth's shape can be gauged from the shadow it casts at the Moon.

Nowadays such proof is not needed. On the other hand, the lunar eclipse enables us to gauge the structure of the upper layers of the Earth's atmosphere by the brilliance and coloring of the Moon. As you know the, Moon does not vanish without trace into the Earth's shadow, it remains visible in the sunrays that curve into the umbra. The strength of moonlight at these particular moments and the Moon's shades of color are of great interest to astronomy, and, it has been established that they are related to the number of sunspots. Furthermore, in recent times lunar eclipse phenomena have been used to measure the rapidity of cooling of the Moon's surface when deprived of the Sun's warmth. (We shall return to this question later.)

Why Do Eclipses Recur Every Eighteen Years

Long before our era, Babylonian star-gazers noted that eclipses, both solar and lunar, recurred every 18 years and 10 days. The ancients named this period the Saros and by means of it were able to predict eclipses. They did not know, however, the reason for this regular pattern of periodicity or for its specific length. This was discovered much later after careful study of the Moon's motions.

What is the time equivalent of the Moon's revolution? The answer varies, depending on what is taken as the completion of the Moon's revolution about the Earth. Astronomers distinguish five kinds of months, of which only two interest us at the moment.

First, the "synodic" month -the time needed for the Moon to make one complete circuit, provided the motion is observed from the Sun. This is the period between two equal phases, for instance, from new Moon to new Moon, and equals 29.5306 days.

Second, the "draconic" month -the time needed for the Moon to return to the same node of its orbit, the node being the point where the lunar orbit intersects the ecliptic. This takes 27.2122 days.

The eclipse, as we can easily comprehend, occurs only when the full or new Moon reaches one of its nodes. Its center will then be on the straight line between the centers of the Earth and the Sun. Clearly, if an eclipse were to take place today, it would recur within an interval containing a whole number of synodic and draconic months, as the conditions for the eclipse would recur.

How do we arrive at these intervals of time? To work them out we must solve the equation $29.5306\,x = 27.2122\,y$ where x and y represent whole numbers. If we turn it into the ratio

$$x/y = 272122/295306,$$

we shall see that the minimum exact solution of this equation is $x = 272122$ and $y = 295306$. Thus we obtain an enormous period of time of the order of tens of millenniums, which is useless from the practical point of view. Ancient astronomers were content with an approximate solution. The continued fraction provides the most convenient means for approximation. Transform the fraction 295306/272122 into a continued fraction as follows. But excluding the whole number we get

$$295306/272122 = 1 + 23184/272122$$

In this last fraction we divide the numerator and the denominator of the by the numerator, thus:

$$295306/272122 = 1 + 1/(1 + 17098/23182)$$

Then we divide the numerator and denominator of the fraction 17098/23182 by the numerator; then again perform a similar operation and so on. Finally we obtain

$$295306/272122 = 1 + 1/(11 + 1/(1 + 1/2 + 1/(1 + 1/(4 + 1/(4 + 1/17 + 1/...)))))$$

Of this fraction, taking its first sections and discarding the rest, we obtain the following consecutive approximations:

$$12/11,\ 13/12,\ 38/35,\ 51/47,\ 242/223,\ 1019/939,\ \text{etc.}$$

The fifth fraction in this progression already yields sufficient accuracy, and if we are to make do with it, that is, accept x as 223 and y as 242, we shall find the periodicity of eclipse recurrence to be 223 synodic months or 242 draconic months, which add up to 6,585 1/3 days, i.e., 18 years, 11.3 (or 10.3[7]) days.

Such is the origin of the Saros. And knowing its origin, we know how exactly we can forecast eclipses. Note that in count-

---

[7] This depends whether the period includes four or five leap years.

Yakov Perelman

ing the Saros as equal to 18 years and 10 days, we discard the odd 0.3 days. This subtraction will be felt in the forecast for an eclipse due at an hour of the day different from the previous eclipse (roughly eight hours later). Only if we take a period thrice the exact Saros will the eclipse recur almost at one and the same time of the day. Moreover, the Saros disregards the changes in distance between the Moon and the Earth and the Earth and the Sun, changes which also have their periodicity, and which determine whether the solar eclipse will be total or not. Thus the Saros enables us to predict merely that there will be eclipse on a certain day. We cannot state definitely whether it will be full, partial, or annular, or whether it can be observed at the same place as its predecessor.

Finally, it happens sometimes that an insignificant partial solar eclipse taking place within 18 years diminishes its phase to naught, making it absolutely unobservable. Conversely, sometimes a hitherto invisible partial solar eclipse becomes observable.

In our times astronomers do not use the Saros. The fickle motions of the Earth's satellite have been studied so thoroughly that eclipses can be forecast to the exact second. Should a predicted eclipse fail to occur, contemporary scholars would be ready to swear the reason was anything you please save an error in calculations. This was very aptly noted by Jules Verne in his *Le pays des fourrures* where he tells the story of an astronomer who set out on a Polar voyage to observe a solar eclipse which, contrary to predictions, did not take, place. What conclusion did our astronomer draw? He informed his companions that the particular ice-field on which they were boated was not mainland but floating ice which the drifts had carried out of the shadow path of the eclipse. This claim was soon proved. A fine example of faith in the might of science!

Can It Happen?

Eye-witnesses say that during lunar eclipses they have seen

the Sun's disc on the horizon on one side and simultaneously the darkened disc of the Moon on the other.

This was also seen during the partial lunar eclipse on July 4, 1936. "The Moon rose at 8.31 p.m. on July 4 and the Sun set at 8.46 p.m. The eclipse took place while the Moon was rising, though both the Moon and the Sun were visible at one and the same time above the horizon. I was greatly surprised at this," a reader of this book wrote to me, "since light rays extend in a straight line."

This, certainly, is a riddle. Although you cannot see through smoke- dimmed glass "the line joining the centre of the Sun and the Moon" it is possible, of course, to imagine it by-passing the Earth in that way. Can an eclipse take place without the Earth obscuring the Moon from the Sun? Can one credit such an eye-witness?

In reality there is nothing incredible about the matter. The Sun and the darkened Moon can be seen in the sky at the same time due to deflection of light as it passes through the Earth's atmosphere. Due to this aberration, known as "atmospheric refraction," we see every celestial object above its true position (Figure 15). When we see the Sun or the Moon on the horizon they are, geometrically, actually below it. So, there is nothing incredible about seeing the Sun and the darkened Moon both on the horizon at one and the same time.

"People usually note," says Flammarion on this score, "the eclipses of 1666, 1668 and 1750 when this peculiar feature was thrown into sharpest relief. But there is no need to go back so far. On February 15, 1877, the Moon rose in Paris at 5:29, and the Sun set at 5:39, although, incidentally, a total eclipse had already begun. December 4, 1880 saw a total lunar eclipse in Paris. On that day the Moon rose at 4:00, and the Sun set at 4:02, almost in the middle of the eclipse, which lasted from 3:03 to 4:33. The reason why this is not seen more often is because few observe it. In order to see totally eclipsed Moon before the setting or after the rising of the Sun, we should choose a point on the globe where the Moon would be on the horizon about the middle of the eclipse."

What Not All Know About Eclipses

Questions

1. How long can a solar eclipse and a lunar eclipse last?

2. How many eclipses are possible in one year?

3. Are there years without a solar eclipse? Or without a lunar eclipse?

4. When will the next total solar eclipse, observable in the U.S.S.R tale place?

5. Does the dark disc of the Moon creep across the Sun from right or left during an eclipse?

6. Does the lunar eclipse begin on right or left?

7. Why do the splashes of light in the shade of leaves take the form of crescents during a solar eclipse? (Figure 59)

8. In what way does the shape of the solar crescent during an eclipse differ from the usual crescent?

9. Why do we need smoke-dimmed classes to observe a solar eclipse?

Answers

1. The longest full phase of a solar eclipse is 7 1/2 minutes (i.e.,

Fig. 59. During a partial eclipse phase, the splashes of light in the shade of leaves take the form of a crescent.

on the equator, being less in higher latitudes). All the phases of the eclipse may last as long as 4 1/2 hours (on the equator).

The phases of the lunar eclipse last up to four hours. The total lunar eclipse never lasts more than 1hr. 50 min.

2. The greatest number is 7, the least 2, both solar and lunar, in one year. (1935 had seven eclipses- 5 solar and two lunar.)

3. Not a year passes without at least two solar eclipses. A year often passes without a lunar eclipse; this happens roughly every 5 years.

4. The next total solar eclipse, observable in the U.S.S.R., will take place on February 15, 1961. The shadow path will cover the Crimea, Stalingrad and Western Siberia.

5. In the Northern Hemisphere the disc of the Moon moves across the Sun from right to left. Its first contact with the Sun should always be anticipated from the right. In the Southern Hemisphere the motion is vice versa (Figure 60).

6. In the Northern Hemisphere the Moon enters the Earth's shadow from the left, in the Southern vice versa.

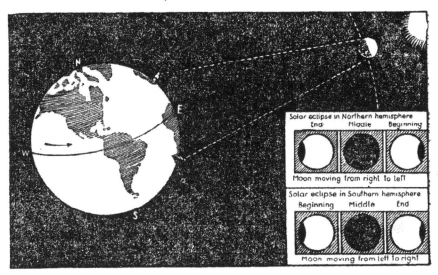

Fig. 60. Why during an eclipse does the observer in the Northern Hemisphere see the Moon's disc creep into the Sun from the right, and the observer in the Southern Hemisphere the reverse?

7. The splashes of light in the shade of leaves simply depict the Sun. During the eclipse the sun is crescent shaped, hence, its image in the shade of the leaves is similarly shaped (Figure 59)

8. The lunar crescent has a semi-circle as its outer rim and a semi- ellipse as its inner rim. The solar crescent lies between two arcs of a circle with one and the same radius. (See section "The Riddle of the Lunar Phases.")

9. We cannot look with the naked eye at the Sun even though partially hidden by the Moon. Its rays burn the most sensitive part of the retina, noticeably depreciating sharpness of vision for a period and, sometimes, for life.

Way back in the early 13th century, a Novgorod chronicler noted: "Many in Great Novgorod lost sight from this celestial portent." We can easily avoid this misfortune by having a well-dimmed glass handy. It should be heavily blurred by candle smoke so that the Sur) is seen as a distinct disc, devoid of rays or halo; for convenience's I sake the smoke-dimmed side can be covered with a clean glass and the edges glued with paper. As we do not know beforehand what the conditions for observing the solar eclipse will be, it will not be amiss to have several glasses of varying dimness ready.

Colored glasses can be used so long as we put together two glasses of different colors (preferably "complementary"). The usual dark eye-glasses are inadequate. Finally, negative photo-plates with appropriately dark enough sections may also be used.

What Is Lunar Weather Like?

Actually the Moon has no weather at all, that is, in our understanding of weather. Indeed, what weather can there be where there is neither air, clouds, vapor, rain nor wind. All we can talk about is surface temperature.

That being so how warm is the surface? Astronomers now have instruments which can measure the temperature not only of remote celestial objects but also of parts of them. These instruments are designed on the principle of thermo-electricity. In a soldered conductor of two heterogeneous metals an electric current is produced when one of the soldered parts is warmer than the other; the power obtained depends on difference in temperature and enables us to gauge the quantity of absorbed warmth.

The instrument's sensitivity is astounding. Despite its microscopic dimensions (the key part being no bigger than, 0.2 mm. and weighing no more than 0.1 mgr.), it reacts even to the warmth of a 13th-magnitude star, which raises the temperature by but the $10,000,000^{th}$ fraction of a degree. These stars cannot be seen without a telescope and their brilliance is but $1/600^{th}$ of the stars on the borderline of naked-eye observation. The effect of this negligible quantity of warmth would be about the same as the heat emitted by la candle several kilometers away.

With this almost miraculous measuring device at hand, astronomers applied it to different parts of the Moon's telescopic image, measured the absorbed warmth and so estimated the temperatures of the different parts with an accuracy of up to 10°. Here are the results (Figure 61): In the centre of the disc of the full Moon, the temperature is more than 100°C; water spilled here would boil even under normal pressure. As one astronomer put it, "we would not need a range took dinner on the moon, any nearby rock would easily suit the purpose." From the centre of the disc the temperature drops evenly in all directions, but even 2,700 km distant from the centre it is not less than 80°C. It then drops rapidly to 50°C. below zero near the rim of the illuminated hemisphere. It is colder still on the dark face the Moon turns away from the Sun. Here it is as much as 160°C below zero.

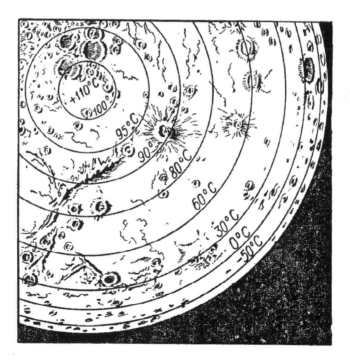

Fig. 61. The temperature in the centre of the Moon's visible hemisphere reaches +110°C., and towards the outer rim falls rapidly to as low as —50°C. and more.

Chapter 2

I noted earlier that during eclipses, when the Moon is sub-merged in the Earth's umbra, its surface, deprived of sunshine, rapidly cools. The cooling was measured and revealed in one case a drop in temperature during the eclipse from 70°C above zero to 117°C below, or by nearly 200°C in 1½ or 2 hours. Meanwhile similar conditions on Earth, i.e., a solar eclipse, produce la temperature drop of only 2°C or, at the most 3° C. This difference is attributed to the Earth's atmosphere which, while comparative-ly transparent for visible sunshine, retards the invisible "heat" rays of the heated surface.

The fact that the Moon's surface loses its accumulated heat so quickly simultaneously indicates its poor heat absorption and conduction capacity, due to which only a small quantity of heat is accumulated.

CHAPTER THREE

THE PLANETS

The Planets in Daylight

Can the planets be seen in daytime during bright sunshine? In the telescope, yes. Astronomers often observe planets in the daytime,, even through medium-powered telescopes, true not so clearly as at night. With a 10 cm. telescope it is possible not only to see Jupiter in the daytime, but even to distinguish its characteristic bands. As for Mercury, it might even be more convenient to observe it in daytime when it is high above the horizon. After sunset Mercury is so low in the sky that the Earth's atmosphere appreciably distorts its telescopic image.

The brightest of them, Venus, is seen most frequently in daylight, naturally, when it is most brilliant. There is the well-known story by Arago about Napoleon who, triumphantly parading through Paris, took affront when the crowd, dumbfounded by Venus' appearance at noon, forsook his exalted person for the star.

The daytime Venus is seen better in the streets of cities than in open places, because the tall buildings obscure the Sun, thus protecting the eye from the dazzling effect of its direct rays. Russian chroniclers have also noted cases of a daytime Venus. Novgorod chronicles say that in 1331, in the daytime "the portent of a bright star did appear in the heavens above the church." The star (as D. O. Sviatsky and M. A. Viliev found later) was Venus.

The most opportune moments for observing the daytime Venus recur every eight years. Attentive star-gazers may have been lucky enough to see with the naked eye in daytime not only Venus but also Jupiter and even Mercury.

This will be the appropriate place to speak about the comparative brilliance of the planets. Laymen sometimes query: Which is more brilliant: Venus, Jupiter, or Mars? Of course, if they shone simultaneously or were placed next to each other, this question would not arise. But as they appear at different times and in different places it is quite a teaser to say which is brighter. The order of brilliance is this: Venus, Mars, and Jupiter are several times brighter than Sirius, while Mercury and Saturn, though weaker than Sirius, are brighter than stars of the first magnitude.

We shall return to this point in the next chapter.

The Planetary Alphabet

In designating the Sun, Moon and the planets, contemporary astronomers employ symbols of a very ancient origin (Figure 62). These symbols need explanation, apart, of course, from that of the Moon which is self-explanatory. The sign for Mercury is a simplified picture of the wand of the legendary god Mercury, the patron of this particular planet. For Venus we have a sign depicting a hand-mirror, the emblem of the femininity and beauty of this goddess. Mars' symbol, as the ward of the god of war, is a spear behind a shield, the arms of the warrior: For Jupiter the sign is simply the first letter of the Greek name for Jupiter, viz., Zeus (the Z is in longhand). According to Flammarion the sign for Saturn is the distorted pictur6 of the "scythes of time," the traditional appurtenance of the god of fate.

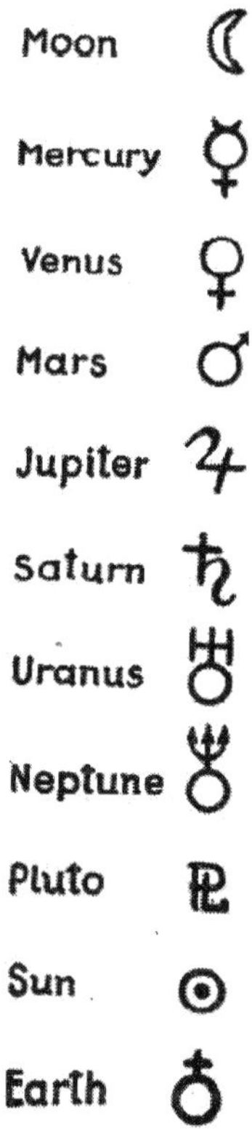

The signs listed above have been used from the 9th century onwards. Uranus' symbol is, naturally, of later origin, since it was discovered only towards the end of the 18th century. Its sign, a circle with the letter H above it, is to remind us of Herschel, who discovered it. The symbol of Neptune, discovered in 1846, pays homage to mythol-

Moon	☾
Mercury	☿
Venus	♀
Mars	♂
Jupiter	♃
Saturn	♄
Uranus	♅
Neptune	♆
Pluto	♇
Sun	☉
Earth	♁

Fig. 62. The symbols designating the Sun, the Moon and the planets.

ogy by depicting the trident of the god of ocean depths. The sign of the last planet, Pluto, is obvious.

To this planetary alphabet we should add the symbol designating the planet on which we live, and the sign for the central body in our system, the Sun. This last symbol holds the palm for antiquity, the Egyptians using it thousands of years ago.

Many will probably think it strange that Western astronomers employ the same symbols of the planetary alphabet to designate the days of the week, namely: the Sun for Sunday, the Moon for Monday, Mars for Tuesday, Mercury for Wednesday, Jupiter for Thursday, Venus for Friday, and Saturn for Saturday.

This association is quite natural if we compare the planetary symbols not with the Russian, but the Latin or French names for the days of the week, names which have preserved their bond with the name of the planets (in French Monday is *Lundi*, the day of the Moon, Tuesday is *Mardi*, the day of Mars, etc) but we shall not elaborate on this curious analogy, as it relates more to philosophy and the history of culture than to astronomy.

Alchemists of yore employed the planetary alphabet to designate metals, using the sign of the Sun for gold, the Moon's for silver, Mercury's for mercury, Venus' for copper, Mars' for iron, Jupiter's for tin, and Saturn's for lead.

This is explained by the mode of thinking of the alchemists who dedicated each metal to one of the ancient mythological divinities.

Finally, we find traces of mediaeval reverence for the planetary symbols in the use of the signs of Mars and Venus by modern botanists and zoologists to designate male and female. Botanists also use the Sun's astronomical symbols to designate annual plants. They employ the same sign in a somewhat modified form (with two dots inside the circle) to designate biennial plants. The Jupiter sign designates perennial grasses. That of

Saturn, bushes and trees.

Something We Cannot Draw

An exact plan of our solar system is something that just cannot be put down on paper. What books on astronomy give as a plan of the solar system is actually a drawing of planetary orbits; it is not at all the solar system; these drawings cannot depict the planets without grossly violating scale. Contrasted with the distance at which they are set apart, the planets are so negligible that it is even hard to have anything like a correct notion of the proportion. We shall find it easier to imagine what it is like if we turn to a diminished likeness of the planetary system. It will then become clear why we cannot pencil the solar system on paper. All we can do is to show the comparative proportions of the planets and the Sun (Figure 63).

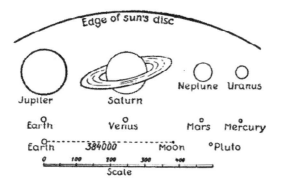

Fig. 63. A comparison of solar and planetary dimensions. On this scale the Sun's diameter is 19 cm.

To depict the Earth we shall choose a most modest value, say, a pinhead. Suppose the Earth is a ball about 1 mm. in diameter. Thus we shall be using a
scale of roughly 15,000 km for 1 mm, or 1:15,000,000,000. The Moon, a tiny speck $^1/4$ mm. in diameter, should be placed 3 cm from the

pinhead. The Sun, a ball 10 cm. in diameter, should be placed 10 m. from the Earth. Thus, a ball in one corner of a spacious room and a pinhead in the other give us an idea of how Sun and Earth are set in relation to each other in space. You can see that here we indeed have fair more space than matter. True, there are two planets between the Sun and the Earth -Mercury and Venus- but they do very little to fill the empty space; all they could add to our room would be two tiny dots, one $^1/_3$ mm. in diameter representing Mercury and placed 4 m. from our ball-Sun, the other for Venus, 7 m. from the ball.

There will also be grains of matter on the other side of the Earth. Mars, $^1/_2$ mm. in diameter, revolves 16 m. from the ball-Sun. Every 15 years the two grains representing Earth and Mars reach their closest conjunction of 4 m apart: this is the shortest distance between these two worlds. Mars has two satellites but we cannot depict them on our model because in the scale we have chosen they would be about the size of a germ. The *asteroids* or minor planets of which some 1,500 rotate in the space between Mars and Jupiter, would have practically the same infinitesimal dimensions. Their mean distance from the Sun would be 28 m. on our scale. The largest of the asteroids would have the thickness of a hair ($^1/_{20}$ mm) on our model, the tiniest, the size of a germ.

We should have to designate the giant Jupiter by a sphere the size of a nut (1 cm), placed 52 m. from the ball-Sun. The biggest of its 12 satellites would circle in its vicinity respectively 3, 4, 7 and 12 cm. away. The diameter of Jupiter's biggest moons would be some $^1/_2$ mm, the remainder again being germ-size. Its most distant satellite, IX, would be 2 m. from the nut representing Jupiter. Hence, the entire Jovian system would be 4 m. in diameter on our model. This is quite a size, compared to the Earth-Moon system with its 6 cm. diameter, but rather modest in contrast to the 104 m. diameter of Jupiter's orbit.

We can see, now how hopeless it is to try to depict the solar system on one drawing. The farther we go, the more striking this becomes. We should have to place Saturn 100 m. from the ball-

Sun, and use a nut, 8 mm. in diameter. The famous Saturnian rings, 4 mm. wide and 1/250 mm thick, would be I mm. away from the surface of this nut. The nine satellites would be scattered about the planet at a distance of half a meter, in the shape of grains with a diameter of 1/10 mm and less.

The empty space between the planets increases in progression the nearer we get to the fringe of the system. Uranus on our model would be 196 m. away from the Sun and, in itself, be a tiny pea 3 mm. in diameter, with five specks for satellites distributed at a distance up to 4 cm from the primary.

The planet, believed until recently the last in our system, namely, Neptune, would be represented by a tiny pea, with two satellites, Triton and Nereide, placed respectively 3 and 70 cm. from it, and would slowly revolve 300 m. away from the central ball.

Still farther away, 400 m. on our model, is the small planet of Pluto, with a diameter about half that of the Earth.

Nor can we reckon the orbit of this last planet to be the boundary of our Solar system, because in addition to planets, it has comets, many of which move in locked paths around the Sun. Among these "longhaired stars" (the, true meaning of the word "comet") there are some with a period about 800 years. These are the comets of 372 B. C., 1106, 1668, 1680, 1843, 1880, f882 (2 comets) and 1887 A. D. On our model the path of each would be an extended ellipse, one end of which, the nearest (the perihelion) would be only 12 mm. from the Sun, while the furthermost point (the aphelion) would be 1,700 m. away, four times the distance of Pluto. Were we to calculate the dimensions of the solar system on the basis of these comets, our model would extend to $3^1/2$ km. in diameter and occupy an area of 9 sq km with, mind you, the Earth no larger than a pinhead! On these 9 sq km we would have the following: one ball, two nuts, two peas, two pinheads, and three smaller specks.

We dismiss the comets, despite their number. Their mass is so small, that they are quite justly called the "visible nothing."

We see, therefore, that our planetary system cannot be drawn on a correct scale.

Why Is There No Atmosphere on Mercury?

What is the connection between the presence of an atmosphere on a planet and the duration of its axial rotation? It would seem there is none at all. Nevertheless, we see from the example of the Sun's nearest planet, Mercury, that there is a connection in some cases.

Mercury's surface gravitation is strong enough to retain an atmosphere akin to that of the Earth, though perhaps not so dense.

The velocity needed completely to overcome Mercury's surface gravitation is 4,900 m. per sec., a speed which even the fastest of molecules in our atmosphere cannot attain at low temperatures.[1] Yet Mercury has no atmosphere. The reason is that it moves about the Sun in the same way as the Moon travels about the Earth. In other words, it always presents one and the same face, to the primary. The time of its orbital passage (88 days) is equal to its axial rotation. Hence on one side of Mercury, that constantly turned to the Sun, there is continuous day and perpetual summer; on the other, that turned away from the Sun, there is continuous night and perpetual winter. One can easily imagine how hot it is on the planet's "daylight" face. Here the Sun is two and a half times nearer than it is to the Earth; consequently the heat effect of its rays should be 2.5 x 2.5, i.e., 6¼ times greater. Conversely, the night bound side, untouched by the Sun's rays for millions of years, must be a frigid zone, with a temperature approximating to the cold of space[2] (around -264° C.), since the

-1- See Chapter 2, "Why Is There No Air on the Moon?"
-2- By "space temperature" physicists mean that indicated in space by a black-shaded thermometer, screened from the Sun's rays. It is somewhat

heat on the "daylight" side cannot penetrate through the planet. On the terminator there is a strip 23° wide, into which, due to libration,[3] the Sun peeps only for a time.

What happens to a planet's atmosphere in such an unusual climate? Evidently due to the severe cold, the atmosphere on the night-bound hemisphere condenses and freezes. The drastic drop in atmospheric pressure will cause the gas enveloping the "daylight" face to rush to the other hemisphere where it freezes. As a result, the entire atmosphere should, in solid form, gather in the night-bound part, or, to be more exact, in the section untouched by the Sun. Thus, Mercury's lack of an atmosphere is the inevitable outcome of physical laws.

For the same reason we must also reject the frequently voiced surmise about the Moon having an atmosphere on its invisible side. We can state quite definitely that if there, is no atmosphere on one side, there cannot he any on the other. Here H. G. Wells' fantastic *The First Men in the Moon* is in error. According to the novelist, the Moon is supposed to have air, which throughout one night of two weeks' duration condenses and freezes, but with the return of day resumes its gaseous state and becomes atmosphere. Actually nothing of the sort takes place. "If," Professor O. D. Khvolson writes in this connection, "the air on the Moon's unlit side solidifies, then nearly all the air should flow from the lit to the dark side and freeze. The Sun's rays would turn the solid air into gas. This gas, in its turn, would immediately flow to the unlit side and solidify there... There would be continuous distillation of air, which could never reach any noticeable elasticity."

While we can affirm that neither Mercury nor the Moon has atmosphere, the contrary is true for Venus, the second planet in distance from the Sun.

---

higher than absolute zero, -273°C., owing to the heating effect of stellar radiation. See Y. Perelman's *Do You Know Your Physics?*
-3- See "The Moon's Visible and Invisible Faces" (Chapter 2). The approximate rule obeyed by the Moon holds good for Mercury's longitudinal librations; Mercury has one hemisphere constantly facing not the Sun but the other focus of its rather extended orbit.

It has also been found that Venus' atmosphere, or to be exact, *stratosphere*, is rich in carbon dioxide, tens of thousands of times more than in the Earth's atmosphere.

The Phases of Venus

The renowned mathematician Gauss relates that he once asked his mother to look at Venus, then shining brightly in the evening heavens, through the telescope. He wanted to surprise his mother, since in the telescope Venus appears as a crescent. But it was Gauss who was startled, for his mother, not the least bit astonished, merely asked why the crescent was turned the other way.... Gauss had no idea at the time that his mother could distinguish the phases of Venus with the naked eye. Such acuteness of vision is rare; and before the spying glass was invented nobody dreamed that Venus had phases akin to those of the Moon.

The peculiar feature of the Venus phases is that at different phases its diameter varies: the narrow crescent diameter is far longer than that of the full disc (Figure 64). The reason for this is the planet's varying distances from us during its different phases. Venus' mean distance from the Sun is 108,000,000 km., that of the Earth-150,000,000 km. A simple reckoning shows that the nearest distance between the two planets is equal to the difference between 150 and 108, i.e., 42,000,000 km., and the farthest distance the sum of 150 and 108, i.e., 258,000,000 km. Hence, Venus' distance from us varies within these limits. When nearest to the Earth, Venus presents its unlit face to us, hence its largest phase is absolutely unobservable. As it moves away from the position of the "new Venus," the planet assumes the form of a crescent, the diameter of which decreases the fuller the crescent.

Venus is most brilliant not when seen as a full disc, nor when its diameter is longest, but at a certain intermediary phase; the full disc is seen at a 10" angle of vision, its biggest crescent at a

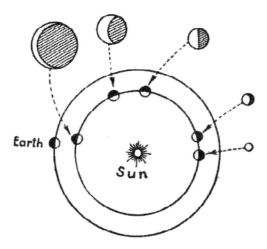

Fig. 64. The phases of Venus as seen through the telescope. Venus' apparent diameter varies at different phases owing to the changing distance from the Earth.

64" angle. On the other hand, the planet is most brilliant 30 days after the "new Venus," when its angular diameter is 40" and angular width of crescent 10". It then shines with brilliancy 13 times that of Sirius, the brightest star in the heavens.

The Most Favorable Opposition

Many know that the times of Mars' greatest brilliancy and closest vicinity to the Earth recur roughly every 15 years. The astronomical name for this, the Most Favorable Opposition of Mars, is exceedingly popular. Red-letter years of recent "oppositions" of this planet are 1924, 1939 (Figure 65) and 1956. But few know why this recurs every 15 years. Incidentally, the related "mathematics" are very simple.

The Earth travels one full orbital passage in 365¼ days, Mars in 687 days. If, say, the two planets once came into conjunction when nearest to one another, they should do so again after an interval consisting of a *whole* number of both terrestrial and Martian years.

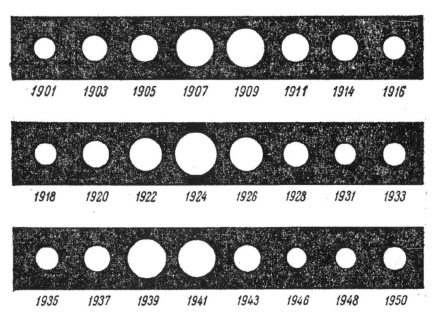

Fig. 65. How Mars' apparent diameter changed during 20th century oppositions. There were most favourable oppositions in 1909, 1924, and 1939.

In other words we must solve in whole numbers the equation

$$356 \ (1/4) \ x = 687 \ y$$

Or

$$x = 1.88 \ y$$

From whence

$$x/y = 1.88 = 47/25$$

Converting the last named fraction into continued one we obtain:

$$\frac{47}{25} = 1 + \cfrac{1}{1 + \cfrac{1}{7 + \cfrac{1}{3}}}$$

The first three sections give us the approximation

$$1 + \cfrac{1}{1 + \cfrac{1}{7}} = \frac{15}{8}$$

and thus we have 15 Earth years equal to 8 Martian years; hence, Mars should be in closest opposition every 15 years. (We have somewhat simplified the problem by accepting 1.88 instead of the more exact 1.8809).

The same method can be used to find the recurrence also of Jupiter's closest oppositions. A Jovian year is equal to 11.86 (11.8622) terrestrial years. Converting this fraction into a continued one we get:

$$11.86 = 11 \frac{43}{50} = 11 + \cfrac{1}{1 + \cfrac{1}{6 + \cfrac{1}{7}}}$$

The first three sections give us the approximation 83/7. Consequently, Jupiter's closest oppositions recur every 83 Earth years, or every seven Jovian years. These are the years when Jupiter, visually, is most brilliant. Its last closest opposition occurred in 1927. The next will come in 2010. Jupiter will then be 587,000,000 km. distant from the Earth -the nearest the solar system's biggest planet ever gets to the Earth.

A Planet or Minor Sun?

That is the question we can ask about Jupiter, largest of planets. The gravitational pull of this giant, which is big enough to make 1,300 Earths, compels a whole swarm of satellites to revolve around it. Astronomers have found that Jupiter has twelve moons, the biggest being the four discovered by Galileo three centuries ago, and designated by the Roman numerals I, II, III, IV. Satellites III and IV are every bit as big as Mercury.

The table below compares the diameters of the satellites with those of Mercury and Mars, and also indicates the diameters of Jupiter's first two satellites and our Moon:

Name	Diameter
Mars	6,600 km
Jupiter's IV satellite	5,150 km
Jupiter's III satellite	5,150 km
Mercury	4,700 km
Jupiter's I satellite	3,700 km
Moon	3,480 km
Jupiter's III satellite	3,220 km

Figure 66 illustrates this table. The large circle is Jupiter, each of the, spheres arranged along its diameter represents the Earth, on the right ere the Moon, Mars and Mercury, and on the left, Jupiter's four biggest satellites. Bear in mind that this is not a diagram but a drawing. The correlation of the areas of these

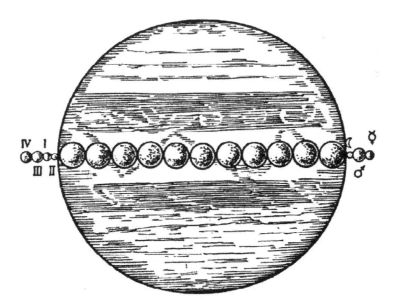

Fig. 66. Jupiter and its satellites (left), compared with the Earth (along the diameter), and the Moon, Mars and Mercury (right).

spheres will not furnish the correct correlation for their *volumes,* which are proportionate to the *cubes* of their diameters.

Since Jupiter's diameter is 11 times bigger than the Earth's, its volume should be 11^3, i.e., 1,300 times more. Correct, accordingly, your visual impression of Figure 66 and you will have a proper notion of Jupiter's immensity.

As for Jupiter's gravitational pull it is truly impressive, especially when you think of the distances at which this giant planet compels its moons to revolve around it. Here is a table of the distances:

Distance	In Km	N-fold
Of the moon from the Earth	380,000	1
Of Io from Jupiter	422,000	1.11
Of Europa from Jupiter	670,900	1.77
Of Ganymede from Jupiter	1,070,000	2.82
Of Callisto from Jupiter	1,883,000	4.96

You will gather from this that the Jovian system is 63 times bigger than the Earth-Moon system; no other planet has such a widely scattered family of satellites.

So it is not without reason that Jupiter is likened to a small sun. Its mass is three times the aggregate mass of all the other planets, and if the Sun were suddenly to vanish its place could well be filled by Jupiter which would compel all the planets to revolve around it, slowly it is true, as the primary.

Jupiter and the Sun have similarities also in physical structure. Jupiter's mean density -1.35 of water- is close to that of the Sun (1.4). However, Jupiter's rather oblate shape gives the impression of dense core enveloped in a thick layer of ice and a vast atmosphere.

Until quite recently the likening of Jupiter to the Sun was

Yakov Perelman

carried farther; it was supposed that it had no solid crust and that it had barely emerged from the phase of a luminary. This view has now been discarded -a direct measurement of Jupiter's temperature showed it to be extremely low, as much as 140° C below zero. True, this refers to the temperature of the clouds floating in its atmosphere.

Jupiter's low temperature greatly complicates the task of explaining its physical peculiarities: its atmospheric storms, bands, spots and so on. Here astronomy is faced with a whole spool of riddles.

Jupiter's atmosphere (and of its neighbor Saturn) was recently found to quite definitely have large quantities of ammonia and methane.[4]

Saturn's Rings Disappear

One day in 1921 the world was startled by a sensational report: Saturn had been dispossessed of its rings! Splinters of the rings, we read, were hurtling through space towards the Sun and would bombard the Earth on their way. Even the date of this catastrophic collision was given....

This story shows how rumors are born. The source of the sensation was simply that in that particular year Saturn's rings had ceased to be visible for a short while, or, as the astronomical ephemeris then said, had "vanished." Hearsay conceived this literally as the physical disappearance, that is, the destruction of the rings, and embellished the event with all the details of a world disaster: hence the fall-out of ring fragments on the Sun and the inevitable collision with the Earth.

The clamour that was raised by this innocent communica-

-4- The methane content is still greater in the atmospheres of the more remote planets of Uranus and, especially, Neptune. In 1944 Saturn's largest satellite, Titan, was found to have a methane atmosphere.

tion of the astronomical almanac concerning the optical disappearance of Saturn's rings! What causes them to disappear? Bear in mind that they are thin, being some thirty kilometers thick, which, compared to their width, is the thickness of a sheet of paper. Therefore when they edge on to the Sun, their upper and lower surfaces are not illuminated and they become invisible. They are invisible again when their rim faces a terrestrial observer.

The rings are inclined at an angle of 27° to the ecliptic, but during the 29 years of the planet's complete orbit they face the Sun and the terrestrial observer edgewise in two opposite points (Figure 67). At two other points, 90° away from the first two, the rings, on the contrary, present to both the Sun and the Earth their widest surface or, as astronomers say, they "open up."

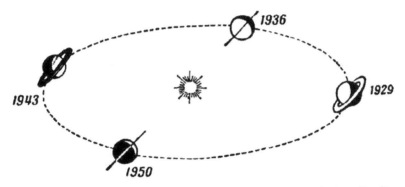

Fig. 67. The positions taken up by Saturn's rings with respect to the Sun within the 29 years of the planet's period of revolution.

Astronomical Anagrams

The disappearance of Saturn's rings strongly perplexed Galileo in his day. He came near to fathoming this remarkable feature but was baffled by the incomprehensible disappearance. The story is a somewhat curious one. In Galileo's day it was customary to confirm priority of the discovery in a rather unusual manner. Upon discovering anything requiring further proof, the scientist or scholar would resort to anagrams (transposition of letters), jealous lest someone should steal a march on him: he would laconically announce the gist of his discovery without undue haste and to claim priority should any pretender appear. The moment he was convinced that his original surmise was right, the scientist would decode his anagram. Upon noting through his imperfect telescope that Saturn has some kind of side appendages, Galileo hastened to announce his discovery and published the following jumble of letters:

Smaismrmielmepoetaleumibuvnenugttaviras.

No one could guess the meaning of the code. One could, of course, try out all transpositions of the 39 letters and thus unravel Galileo's hidden phrase. But, what a job! Anyone familiar with the theory of combinations will be able to express the total of possible transpositions (with repetitions). Here it is:

$$39! / (3! \; 5! \; 4! \; 4! \; 2! \; 2! \; 5! \; 3! \; 3! \; 2! \; 2!)$$

This figure consists roughly of 35 numerals. Bear in mind that the number of seconds in a year amounts "only" to 8 numerals! Now you will appreciate how closely Galileo guarded the secret of his claim.

With the patience for which he was renowned, Kepler, a contemporary of the Italian scientist, labored hard to probe into the hidden sanctuary of Galileo's claim. He believed he had accomplished this task when from the published letters (with two omitted) he formed the following Latin phrase:

Salve, umbistineum geminatum Martia proles
(I salute you, twins born of Mars.)

Kepler was convinced that Galileo had discovered Mars' two satellites whose existence he himself had suspected.[5] (They were

-5- Apparently in this instance Kepler was guided by a surmised progression in the number of planetary satellites. Knowing that the Earth had one satellite and Jupiter four, he believed it quite natural for Mars as the intermediary planet to have two satellites. This line of reasoning led others to think that

actually identified 250 years later). However, this time, Kepler's guesswork misfired. When Galileo finally disclosed his secret, the phrase proved to be (with two letters omitted);

Altissimum planetam tergeminum observavi

(I observed a very high and triple planet.)

Since his telescope was so weak, Galileo could not realize what this "triple" image of Saturn really implied. A few years later when the planet's side appendages disappeared altogether, Galileo decided that he had erred and that Saturn had no appendages at all.

Then, after half a century had passed, Huygens was fortunate enough to discover Saturn's rings. Like Galileo, he did not make-his discovery known at once, but coded his surmise in the cryptogram:

Aaaaaaaccccccdeeeeeghtitutilllllmmnnnnnnnnn

oooooppqrrstttuuuuu

Three years later, convinced that he was right, Huygens published his secret:

Annulo cingitur, tenui, plano, nusquam cohaerente, ad eclipticam inclinato

(Hemmed by circle, thin, flat, nowhere cohering and inclined to ecliptic.)

Mars had two satellites. In Voltaire's astronomical fantasy *Micromegas* (1750) we read of the voyager who upon approaching Mars saw "two moons which served this planet and which had hitherto been hidden from the eye of our astronomers." *Swift's Gulliver's Travels*, written still earlier, in 1726, contains a similar remark, according to which Laputian astronomers had "discovered two lesser stars, or satellites, which revolve about Mars." These curious surmises were fully corroborated in 1877 when Hall: discovered Mars' two satellites.

Trans-Neptunian Planet

In 1929 when this book first appeared I wrote that Neptune was the most distant of the, known planets of the solar system -30 times farther away from the Sun than the Earth. Now I can no longer say so, because 1930 added to the solar system a new member, a ninth major planet, revolving about the Sun still farther away than Neptune.

This discovery did not come as a bolt from the blue. Astronomers had long toyed with the idea of an unknown trans-Neptunian planet.

Just slightly over a century ago they believed Uranus to be the solar system's extreme planet. But certain irregularities in its motions led them to suspect the existence of a planet still farther away whose gravitation upset the even tenor of Uranus' way. Mathematical investigation of the question by the British mathematician Adams and the French astronomer Leverier culminated in a brilliant discovery; the suspected planet was seen in the telescope. The human eye spied a world the existence of which had been established by "penible" calculations.

Such is the story of Neptune's discovery. It was subsequently found that its influence did not explain everything about the irregularity of Uranus' motions. It was conjectured that there was, possibly, another planet still farther away than Neptune. So the mathematicians began to rack their brains. Various solutions were proposed; the supposed 9th planet was placed at varying distances from the Sun and invested with different masses.

In 1930, or to be exact at the end of 1929, the telescope finally extracted from the gloomy murkiness on the fringes of the solar system another member of our planetary Family. The new planet, named Pluto, was discovered by the young astronomer Tombaugh.

Pluto rotates about a path very close to one previously computed. But this, according to the experts, cannot be hailed as a mathematical, success; the coincidence is merely a curious accident.

What do we know about this newly discovered world? So far very little; it is so far away and so meagerly lit by the Sun that even most powerful instruments proved hardly able to measure its diameter, which was found to be 5,900 km or 0.47 of the Earth's diameter.

Pluto revolves around the Sun along a rather extended orbit with an eccentricity of 0.25. It is noticeably tilted at 17° to the ecliptic and is 40 times farther from the Sun than the Earth. It takes the planet 250 years to complete this huge revolution.

In Pluto's heavens the Sun shines with a brilliance 1,600 times fainter than with us and is seen as a tiny disc of 45 angular seconds, i.e., approximately the same as we see Jupiter. It would be interesting to establish, though, which is the more brilliant, the Sun for Pluto or the full Moon for the Earth.

It seems that faraway Pluto is by no means so badly off for sunshine as one would think. The light we get from the full Moon is 440,000 times weaker than that from the Sun. In Pluto's sky our diurnal luminary emits a light 1,600 times weaker than it does for us. Hence, the brilliance of Pluto's sunshine is 440,000/1,600 or 275 times stronger than full moonlight on Earth. If Pluto's skies are as clear as the Earth's, which seems to be the case, since Pluto apparently has no atmosphere, daylight there would be the same as the light of 275 full moons shining at once, or about 30 times brighter than the whitest of white nights in Leningrad. It would be wrong, therefore, to describe Pluto as the kingdom of eternal night.

Pigmy Planets

The nine major planets mentioned so far do not exhaust the planetary population of the solar system. They just happen to be the biggest. In addition, circling about the Sun at varying distances are a legion of tiny planets. These pigmies in the world of planets are known as asteroids (meaning "star-like"), or simply "minor planets." The biggest, Ceres, is 770 km in diameter; it is much smaller than the Moon -about the same number of times that the Moon is smaller than the Earth.

Ceres, first of minor planets, was discovered on January 1, 1801. More than 400 of the pigmies had been spotted during the XIX century. Until recently it was believed that the asteroids were banded together in the wide gap between the orbits of Mars and Jupiter.

The 20th century, recent years, especially, has extended the boundaries of the asteroid belt in both directions. Eros, discovered towards the end of the last century -1898- already broke, these boundaries, as most of its path lay within Mars' orbit. In 1920 astronomers espied the asteroid Hidalgo, whose path intersects Jupiter's orbit and is close to Saturn's. Hidalgo is also remarkable for having the most elongated orbit (eccentricity 0.66) of all known planets and, moreover, is the most strongly tilted to the ecliptic (at 43°).

We might note in passing that this asteroid takes its name from Hidalgo y Costilla, hero of Mexico's independence movement, who was shot in 1811.

In 1936, when an asteroid with an eccentricity of 0.78 was observed, the zone of pigmy planets was found to be still wider. This new member of our solar system was named Adonis and is remarkable for the fact that at its perihelion it is almost as far away from the Sun as Jupiter, while at aphelion it is in the vicinity of Mercury's orbit.

Finally the minor planet Icarus, discovered in 1949, has an exceptional path with an eccentricity of 0.83, a perihelion point twice the radius of the Earth's orbit, and aphelion at about a fifth of the distance between the Sun and Earth. None of the known planets gets as close to the Sun as Icarus.

The system of recording newly discovered asteroids is not without interest -it can be successfully applied for purposes other than astronomical. First, the year of discovery is registered, and then the letter designating the half-months at the time of discovery (the year is divided in 24 half-months, successively designated by letters of the alphabet).

Since several minor planets are often discovered in one half-month, they are designated by one more letter in alphabetical order. Should the 24 letters not suffice, they are repeated but this time with tiny numerals attached. Thus, for instance, 1932 EA_1 signifies an asteroid discovered in the first half of March 1932, the twenty-fifth in that period. When the orbit of the newly discovered planet has been computed, it is given an ordinal numeral and a name.

Of the multitudinous minor planets probably only a small number are as yet accessible to astronomical instruments. According to calculations there must be from 40 to 50 thousand asteroids in the solar system.

The number of pigmy planets recorded to date exceeds 1,500. Over a hundred of these were found by astronomers of the Simeiz Observatory in the Crimea, chiefly due to the diligence Of G. N. Neuymin, an assiduous asteroid hunter. The reader will not be surprised to find in the list of minor planets names like "Vladilen" (in honor of Vladimir Ilyich Lenin), "Morozovia" and "Figneria" (in honor of two Russian revolutionary heroes), "Simeiza" and others. For the number of discovered asteroids Simeiz occupies a leading place among world observatories; and in the elaboration of theoretical questions pertaining to asteroids Soviet astronomers are also prominent. The Institute of Theoretical Astronomy in Leningrad has devoted years of study to pre-charting the positions of a large number of minor planets and elaborating the theory of their motions. Every year it publishes the pre- plotted positions of the minor planets (the so-called *ephemeris*) for observatories all over the world.

The minor planets are exceedingly varied in size. Only a few are as big as Ceres or Pallas (diameter = 490 km). About seventy have diameters of more than 100 km. Most of them have diameters ranging between 20 and 40 km. Then there are the numerous quite "tiny" asteroids with diameters barely 2 or 3 km. (we have put "tiny" in quotation marks, for coming from an astronomer's lips it must be understood relatively). Although far from all the members of the asteroid belt have been charted we may presume the aggregate mass of asteroids (discovered and undiscovered) to be about a thousandth of the Earth's. So far only an estimated five per cent of the asteroids accessible to modern telescopes have been discovered.

"One might think that the physical properties of the asteroids would be roughly the same," G. N. Neuymin, our leading minor-planet expert, wrote. "Actually we encounter a variety that is staggering. For example, determination of the reflection properties of the first four asteroids showed that Ceres and Pallas reflect light in the same way as the Earth's dark rocks, that Juno does so like light rock, and Vesta like white clouds. This is all the more puzzling because, being small, they cannot retain an atmosphere. That has been definitely established, and so we are forced to ascribe the different reflection properties to their surface materials."

Some of the minor planets display a fluctuating brilliance, testifying to axial rotation and irregular shape.

Our Nearest Neighbors

Adonis, the asteroid mentioned above, stands out not only for its exceedingly elongated cornet-like orbit; another feature is that it comes fairly close to the Earth. The year of its discovery found Adonis just 1,500,000 km. away. True, the Moon is nearer, but, while much bigger than the asteroids, it is, nevertheless, of inferior rank, being a satellite, not an independent planet. Apollo is another asteroid that can claim to be one of the Earth's nearest neighbors. The year it was discovered Apollo was a mere 3,000,000 km. away from our Earth. In planetary terms this is no great distance, since Mars, at its nearest, is 55,000,000 km. away, while Venus is never less than 40,000,000 km. Curiously, this same asteroid travels still closer to Venus, to within 200,000 km, half the distance between the Moon and the Earth! So far this is the closest interplanetary relationship known in the solar system.

This planetary neighbor of ours is also remarkable for being one of the smallest ever recorded. Its diameter is certainly not more than 2 km. and may be even less. The year 1937 brought the discovery of Hermes, an asteroid which at times approaches the Earth at the Moon's distance (500,000 km); its diameter is not more than 1 km.

With the foregoing as an example let us look closer at the astronomical meaning of the word "tiny." This tiny asteroid, a mere 0.52 km^3 in volume, i.e., 520,000,000 cu. m., would weigh about 1,500,000,000 tons were it composed of granite. With such a quantity of granite it would be possible to build 300 pyramids like that of Cheops.

That gives you an idea of the word "tiny" when used by the astronomer.

Jupiter's Fellow-Travelers

Among the 1,600 known asteroids, fifteen, named after the Trojan heroes, *Achilles, Patroclus, Hector, Nestor, Priamus, Agamemnon*, etc., stand out for their remarkable motions. Each of the "Trojans" rotates near the Sun so that in any position the three -asteroid, Jupiter and Sun- form the apexes of an equilateral triangle. The "Trojans" can be regarded Jupiter's fellow-travelers, accompanying him at a considerable distance -some up to 60° in front of Jupiter, others just as far behind- but all revolving around the Sun in one and the same period of time.

The balance of the planetary triangle is stable; should the asteroid quit its position, gravitation would bring it back.

Long before the "Trojans" were discovered, a similar mobile balance of three attracting bodies was predicted in the purely theoretical investigation carried out by the French mathematician Lagrange. Lagrange examined this case as a curious mathematical problem and scouted the idea of a relationship of this kind being found anywhere in the universe. However, the assiduous quest for asteroids resulted in a tangible illustration of Lagrange's hypothesis being within the confines of our own planetary system. This is striking proof of the importance of meticulous study of the host of celestial bodies, known as the minor planets.

Alien Skies

We have already made an imaginary flight to the Moon to see what the Earth and other celestial bodies look like from there.

Let us now pay a visit to the planets of the solar system to see what picture of the heavens we obtain from each.

We shall begin with *Venus*. If the atmosphere there were sufficiently transparent, we would see a Sun twice the size of the one shining in our sky (Figure 68). This Sun, then, gives Venus double the Earth's warmth and light. In the nocturnal heavens of Venus we would see a star of unusual brilliance. This would be our Earth, shining far brighter than Venus does in our heavens, though in size the two planets are almost identical. The reason is simple. Venus is nearer to the Sun than the Earth. Therefore when it is nearer to the Earth we do not see it at all because its unlit face is presented to us. To be visible it would have to slip aside somewhat. Light would then be emitted by a narrow crescent, comprising but a small part of Venus' disc. On the other hand, the Earth, when nearest to Venus, shines in its heavens as a full disc, as Mars does in our heavens when in opposition. As a result, the Earth in its full phase in Venus' heavens would be six times brighter than Venus is in our sky during its greatest brilliancy. This should be so, of course, if our neighbor's sky were clear enough. We would err, however, if we imagined that abundant "Earthshine" on a Venus night produced its silvery ash-grey light; the illumination of Venus by the Earth is about the same as that shed by an ordinary candle 35 m away. That, of

Fig. 68. The apparent dimensions of the Sun seen from the Earth and the other planets.

course, is not enough to produce Venus' silvery color.

The "Earthshine" in Venus' heavens is often complemented by the light of the Moon which here is four times brighter than Sirius. It is doubtful if we would find in the entire solar system an object as brilliant as the double, Earth-Moon luminary adorning the heavens of Venus. Most of the time the observer on Venus would see the Earth and the Moon separately, and he would be able to distinguish details of lunar surface through the telescope.

Mercury is another brilliant planet in Venus' sky -its morning evening star. Mercury, incidentally, is also seen from the Earth as a bright star, even outshining Sirius. In Venus' sky it shines with nearly thrice the brightness of that in the Earth's heavens. On the other hand, Mars gives but a fifth of its light, somewhat dimmer than Jupiter in our skies.

As for the fixed stars, the outlines of the constellations are absolutely the same, for the skies of all the planets of the solar, system. Whether we were on Mercury, Jupiter, Saturn, Neptune, or Pluto we would see one and the same stellar tracery. This shows how far away the stars are compared with interplanetary distances.

**

Let us leave Venus for tiny *Mercury,* for a strange world devoid of atmosphere and knowing no alternation of night and day. Here the Sun hangs immobile in the sky, a huge ball six times bigger (in area) than when seen from the Earth (Figure 68). From Mercury our planet is seen as a star shining with twice the brilliance of Venus in our sky; Venus, too, is unusually brilliant here. In fact nowhere else in our system does star or planet shine with the, dazzling brightness of Venus against the black and cloudless sky of Mercury.

Our next stop is *Mars.* Here the Sun is seen as a ball two-thirds less in area (Figure 68). In Mars' heavens our own globe shines as morning and evening star, the proxy for Venus in our sky, but fainter, much as we see Jupiter. Here the Earth is never seen in its full phase; at any one time the Martian would never see more than 3/4 of its disc. He would see the Moon as a naked-eye star about as bright as Sirius. In the telescope both the Earth and its attendant Moon would show their phases.

Here attention would be focused on Mars' nearest satellite, Phobos. So near is it to Mars that, despite its insignificant size (16 km in diameter), the "full Phobos" shines with 25 times the brightness of Venus in our sky. Deimos, the second satellite, while much less brilliant, also outshines the Earth in Mars' heavens. Despite its smallness Phobos is so close to Mars that its phases are clearly seen. Anyone with good eyesight would probably observe even its phases (Deimos is seen from Mars at an angle of 1', Phobos at an angle of nearly 6').

Before going farther let us halt for a moment on the surface of Mars' nearest satellite. From this vantage post we would see the unique spectacle of a giant disc, rapidly changing its phases, several thousand times brighter than our Moon. This is Mars. Its disc takes up 41° of the skies -80 times more than the Moon in our heavens. Only on Jupiter's nearest satellite it is possible to observe a similar unusual but remarkable phenomenon.

* * *

Our next stop is the giant planet we have just mentioned. If Jupiter's skies were clear the Sun would be seen as a sphere 25 times less in area than in our heavens (Figure 68); it would be the same number of times fainter. The brief five-hour day here quickly gives way to night, so let us seek the familiar planets in the starry sky. We shall find them no doubt, but how changed! Mercury is absolutely lost in the Sun's rays, while Venus and the Earth can be seen telescopically only at twilight -they set together with the Sun.[6] Mars is hardly noticeable. On the other hand Saturn successfully rivals Sirius in brilliance.

In Jupiter's heavens prominence is taken by its moons. Satellites I and II are about as bright as the Earth in the skies of Venus, III is three times brighter, while IV and V are several times brighter than Sirius. As for their dimensions, the visual diameters of the first four satellites are larger than the visual diameter of the Sun. At each revolution the first three satellites are plunged in Jupiter's shadow, so that they are never seen in full

-6- In Jupiter's heavens the Earth is seen as a star of the 8th magnitude.

phase. Total solar eclipses are also seen, but can be observed only in a narrow strip across Jupiter's surface.

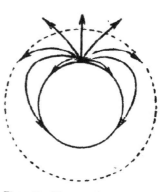

Incidentally, Jupiter's atmosphere could scarcely be so transparent as that of the Earth; it is too high and dense. The considerable density would tend to engender very peculiar optical illusions connected with light refraction. Atmospheric refraction of light on the Earth is insignificant and merely engenders elevation (optically) of the heavenly bodies. On the

Fig. 69. How light probably curves in Jupiter's atmosphere (see text for the consequences of this phenomenon).

other hand, Jupiter's high and dense atmosphere is conducive to far more noticeable optical illusions. Rays emanating obliquely from a point on the surface (Figure 69) do not leave the atmosphere at all, they curve towards the planet's surface, like the radio waves in the Earth's atmosphere. The observer at this point would see something utterly unusual. He would imagine himself at the bottom of a huge bowl. Practically the entire surface of the vast planet would be inside, with the outlines near the rim strongly compressed. Overhead -the sky, not just half of it as is the case with us, but nearly all of it, and ending in a hazy and misty fringe only at the edge of the bowl. Our diurnal luminary would never depart from this strange sky, with the result that the midnight Sun would be visible at all points on the planet. We cannot, of course, definitely say whether Jupiter really has these extraordinary features.

A close-up of Jupiter from its nearest satellite (Figure 70) would present an astonishing sight. Seen from the fifth and nearest satellite its huge disc would have a diameter nearly 90 times bigger than the Moon[7] and would shine with a brightness only six or seven times less than the Sun. When its lower limb touches the horizon, its upper limb is in the centre of the heavens, and when it dips below the horizon it takes up an 8th of the sky-line. <u>From time to t</u>ime dark circles, the shadows thrown by Jupiter's

-7- Seen from the satellite, Jupiter's angular diameter is more than 44°.

Fig. 70. Jupiter as seen from his III satellite.

moons, which are powerless, of course, to "eclipse" this huge planet to any noticeable degree, glide across the rapidly revolving disc.

* * *

Now for a visit to *Saturn* merely to see how its famous rings appear to the observer. We discover firstly that not everywhere on Saturn are the rings visible. For instance, they would not be

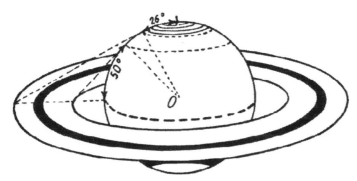

Fig. 71. How we find the degree of visibility of the Saturnian rings for various places on that planet. Between the Pole and the 64th parallel the rings are not seen at all.

seen at all between the Poles and the 64th parallel. On the boundary of this area we would see only the external rim of the outer ring (Figure 71). The view between the 64th and the 50th parallel is better, and on the 50th parallel the observer can admire the entire range of the rings, seen here at their greatest angle of 12°. Nearer to the equator they narrow, though rising higher above the horizon. At the equator itself the rings are seen only as a narrow strip cutting across the zenith from west to east.

But that is not the whole picture. Bear in mind that only one side of the rings is lit, the other being in shadow. The illuminated side is seen only from that half of Saturn which it faces. For half the long Saturnian year we see the rings only from one half of the planet (they are seen from the other half the rest of the time), and even then primarily in the daytime. During the brief hours when the rings are seen at night, they are often eclipsed by the planet's shadow. Lastly, one more curious detail: the equatorial regions are eclipsed by the rings for a number of terrestrial years.

Of all celestial pictures the most spectacular by far is the sight the observer would obtain from one of Saturn's nearest satellites. Seen especially during an incomplete phase, when it has the shape of a crescent, Saturn with its rings is a sight which cannot be seen anywhere else in the planetary system. We would see in the sky an enormous crescent intersected by a narrow band of rings, seen rim wise, and around them a group of Saturnian satellites, also in crescent shape, only much smaller.

* * *

The following list shows, in diminishing order, the comparative brilliancy of the different bodies in the skies of other planets:

1. *Venus from Mercury*

2. *The Earth from Venus*

3.   *The Earth from Mercury*

0.   *Venus from the Earth*

4.   *Venus from Mass*

5.   *Jupiter from Mars*

6.   *Mars from the Earth*

8.   *Mercury from Venus*

9.   *The Earth from Mars*

10.   *Jupiter from the Earth*

0.   *Jupiter from Venus*

1.   *Jupiter from Mercury*

2.   *Saturn from Jupiter*

The Planetary System in Figures

Dimensions. Mass. Density. Satellites

Name	Mean diameter			Volume (Earth =1)	Mass (Earth =1)	Density		Number of satellites
	Visible in "	True				Earth = 1	Water = 1	
		in km.	Earth = 1					
Mercury	13‑4.7	4,700	0.37	0.050	0.054	1.00	5.5	-
Venus	64‑10	12,400	0.97	0.90	0.814	0.92	5.1	-
Earth	-	12,757	1.0	1.00	1.000	1.00	5.52	1

Mars	25-3.5	6,600	0.52	0.14	0.107	0.74	4.1	2
Jupi-ter	50-30.6	142,000	11.2	1,295	318.4	0.24	1.35	12
Sat-urn	20.5-15	120,000	9.5	745	95.2	0.13	0.71	9
Ura-nus	4.2-3,4	51,000	4.0	63	14.6	0.23	1.30	5
Nep-tune	2.4-2.2	55,000	4.3	78	17.3	0.22	1.20	2
Pluto	0.2?	5,900	0.47	0.1	?	?	?	?

The Planetary System in Figures

Distance. Revolution. Rotation. Gravity

Name	Mean distance		Eccentricity of orbit	Time of revolution around the Sun in terrestrial years	Mean orbital velocity in km. per sec.	Period of axial rotation	Tilt of equator to orbital plane	Gravity (Earth =1)
	In a. u.	In million km.						
Mer-cury	0.387	57.9	0.21	0.24	47.8	88d	?	0.26
Venus	0.723	108.1	0.007	0.62	35	30d?	?	0.90
Earth	1.000	149.5	0.017	1.00	29.76	23h56m	23°27'	1.00
Mars	1.524	227.8	0.093	1.88	24	24h37m	25°10'	0.37
Jupiter	5.203	777.8	0.048	11.86	13	9h55m	3°01'	2.64
Saturn	9.539	1,426.1	0.056	29.46	9.6	10h14m	26°45'	1.13
Uranus	19.191	2,869.1	0.047	84.02	6.8	10h48m	98°00'	0.84
Nep-tune	30.071	4,495.7	0.009	164.80	5.4	15h48m	29°36'	1.14
Pluto	39.458	5,899.1	0.250	247.70	4.7	?	?	?

We single out Nos. 4, 7 and 10, or how the planets are seen from our Earth, because being familiar with their brilliance we can use them as a guide in assessing the visibility of luminaries from other planets. Here we can see particularly clearly that for brilliance our own planet occupies a leading place among those nearest to the Sun: even in Mercury's sky it is brighter than Venus and Jupiter are in our sky.

In the section headed "Stellar Magnitude of Planets" (Chapter IV) we shall return to a more exact evaluation of the quantitative brilliance of the Earth and the other planets.

Lastly, a few figures about the solar system, which might come in handy for reference.

The Sun: 1,390,600 km in diameter, 1,301,200 times the Earth in volume, 333,434 times the Earth in mass and 1.41 times in water density.

The Moon: 3,473 km. in diameter, 0.0203 times the Earth in volume, 0.0123 times the Earth's mass, and 3.34 times in water density. Mean distance from the Earth - 384,400 km.

The tables above give the figures for the planets of the solar system.

Figure 72 is a graphic illustration of what the planets would look like through a small telescope, magnified 100 times. On the left is the Moon magnified to the same degree and included for the sake of comparison. (The drawing must be held at a distance of clear vision, i.e., 25 cm. from the eye.) Mercury is shown in the upper right corner, in the given magnification, at its nearest and most distant points from us. Venus, Mars, the Jovian family, and Saturn with its biggest satellites follow in that order. (For greater detail about visual planetary dimensions see my *Physics for Entertainment*, Volume 2, chapter IX.)

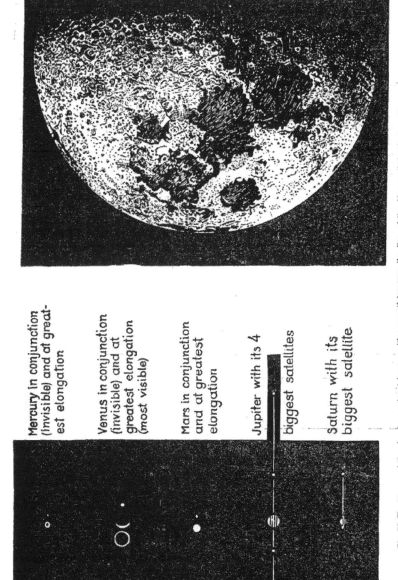

Mercury in conjunction (invisible) and at greatest elongation

Venus in conjunction (invisible) and at greatest elongation (most visible)

Mars in conjunction and at greatest elongation

Jupiter with its 4 biggest satellites

Saturn with its biggest satellite

Fig. 72. The Moon and the planets seen in a telescope with a magnifying eyes; the discs of the Moon and the planets will then be seen as they appear power of 100. The drawing should be held 25 cm. away from the in a telescope of the given magnifying power.

Chapter 3

CHAPTER FOUR

THE STARS

Why Do Stars Look Like Stars?

When we look at the stars with the naked eye we see them emit a star-shaped glitter.

The reason for this is to be found in our own eye, or rather in the inadequate transparency of the crystalline lens, which, unlike a good glass lens, is fibrous in structure, not homogeneous. Here is what Helmholtz (in his *Achievements of the Theory of Vision*) has to say about the matter:

"Points of light produce in the eye an incorrect star-like image. This is due to the crystalline lens, the fibers of which stretch out tentacle-like in six directions. The rays which we take to be coming from points of light, say, the stars or remote lights, are simply the reflection of the radial structure of the crystalline lens. The universality of this shortcoming of the eye is evident from the fact that any ray-like figure is usually said to be star-shaped."

We can, if we wish, remedy this failing of our crystalline lens and see the stars without their radial glitter and, moreover, without using the telescope. Leonardo da Vinci told us how to do it 400 years ago: "Behold the stars without rays," he wrote. "We can do so by looking at them through a tiny aperture, pricked by a fine needle, and brought close to the eye. The stars will appear so tiny that it would seem nothing could be smaller."

This in no way contradicts Helmholtz ideas about the origin of the stellar rays. On the contrary, the experiment described confirms his theory; looking through a tiny aperture the eye takes in only a slender beam of light, which, passing through the central part of the crystalline lens, is not therefore affected by its radial structure.[1]

Thus, if our eyes had more perfect structural qualities we would see in the sky not "stars," but points of light.

Why Do Stars Twinkle While Planets Shine Steadily?

We easily distinguish with the naked eye between a fixed and "wandering" star or planet,[2] even without any knowledge of the celestial atlas. The planets emit a *motionless* shine, while the stars *twinkle* continuously -they seem to flash, tremble and change in brilliance, while the radiant stars, low on the horizon, scintillate endlessly. "This light," says Flammarion, "now bright, now faint, twinkling now white, now green, now red, and sparkling like the purest of diamonds, enlivens the stellar expanses, so that we take the stars for eyes gazing at our Earth." The stars twinkle with particular force and beauty on frosty nights, in windy weather and also after heavy rain when the skies are rapidly cleared of clouds.[3]

The stars on the skyline twinkle much more than those shining high in the sky, with the whitish stars taking priority over those yellow or reddish in hue.

The twinkle, like the star-shaped radiance, is not an intrinsic quality of the stars; it is imparted to them by the Earth's atmosphere through which the rays pass before reaching the eye. If we could raise ourselves above the restless gaseous mass surrounding us, through which we View the universe, we would see that the stars do not twinkle, that they shine with a calm and constant light.

-1- Speaking of "stellar rays" we have in mind not the ray which seems to come from the stars when we screw up our eyes to look at them; this is caused by diffraction of light on the eyelids.

-2- The original meaning of the Greek word "planet" is "wandering star."

-3- Strong summer twinkling is a sign of rain, indicating the approach of a cyclone. Before rain the stars emit primarily a blue light; they shed a green light before drought.

The stars twinkle for the same reason that distant objects tremble during a heat-wave when the ground becomes heated owing to the action of the Sun.

The star-light, then, has to penetrate not a homogeneous medium, but gas layers of diverse temperature and density and, hence, of diverse refractivity. This kind of atmosphere seems to consist of a host of optical prisms, of concave and convex lens, all in constant movement. In passing through them light diverges again and again from a straight course, converges and then disperses again. This is the reason for the frequent change in the brilliancy of stars. And since dissipation of color accompanies the refraction, the result is, in addition to variations in brilliance, a change in hue.

The Pulkovo astronomer G. A. Tikhov, who has studied star twinkling, says: "There are ways and means of reckoning the number of times a twinkling star changes color in a definite period. The changes are effected with the utmost rapidity and vary in number from a few dozen to a hundred and more per second. We can see this is so by resorting to the following simple method. Take a pair of binoculars and look at a bright star, revolving the lens quickly. Instead of the star you will see a ring of numerous multi-colored stars. When twinkling is slow, or when the lens is turned quickly, the ring disintegrates not into stars but into multi-colored arcs of greater or lesser length."

I shall now explain why the planets, in contrast to the stars, do not twinkle but shine evenly and steadily.

The planets are much nearer to us than the stars; hence we see them not as points but as circles or discs of light, and their tiny angular dimensions, thanks to their radiant brilliance, are scarcely noticeable. Every point in this circle of light twinkles, but since the radiance and hue of each change independently and at different times, they complement one another; the fading brilliancy of one point dovetails with the growing brilliancy of another so that the total light shed by the planet remains constant.

Hence, the even unbroken brilliancy of the planets.

So the planets appear to us as non-twinklers because while they twinkle simultaneously at many points, they do so at different times.

Can Stars Be Seen in Daylight?

The constellations that we saw at night half a year ago are now overhead in the daytime. Six months later they will again adorn the night sky. The sunlit atmosphere of the Earth screens them from the eye because the air particles disperse the sunrays more than the rays emitted by the stars.[4]

The following simple experiment will help explain why the stars disappear in daylight. Punch a few holes in one of the sides of a cardboard box, taking care, however, to make them resemble a familiar constellation. Having done so, glue a sheet of white paper on the outside. Place a light inside the box and take it into a dark room; lit from the inside; the holes, representing stars in the night sky, are clearly seen. But, switch on a light in the room without extinguishing the light in the box and, lo, the artificial stars on our sheet of paper vanish without trace: "daylight" has extinguished them.

One often reads of stars being seen even in daylight from the bottom of deep mines and wells, of tall chimney-stacks and so on. Recently, however, this viewpoint, which had the backing of eminent names, was put to test and found wanting.

As a matter of fact, none of the men who wrote on this subject, whether the Aristotle of antiquity or 19th-century Herschel, had ever bothered to observe the stars in these conditions. They quoted the testimony of a third person. But the un-wisdom of re-

-4- The observer located on the top of a high mountain, with the densest and dustiest layers of the atmosphere below, would see the brighter stars even in daytime. For instance, from the top of Mt. Ararat (5 km high), first-magnitude stars are clearly distinguished at 2 o'clock in the afternoon; the sky is seen as having a dark blue color.

lying on the testimony of "eye-witnesses," say in this particular field, is emphasized by the following example. An article in an American magazine described daylight visibility of stars from the bottom of a well as a fable. This was hotly contested by a farmer who claimed that he had seen Capella and Algol in daytime from the floor of a 20-meter high silo. But when his claim was checked it was found that on the latitude of his farm neither of the stars was at zenith at the given date and, consequently, could not have been seen from the bottom of the silo.

Theoretically, there is no reason why a mine or a well should help in daylight observation of stars. We have already mentioned that the stars are not seen in daytime because sunlight extinguishes them. This holds also for the eye of the observer at the bottom of a mine. All that is subtracted in this case is the light from the sides. All the particles in the layer of air above the surface of the mine continue to give off light and, consequently, bar the stars to vision.

What is of importance here is that the walls of the well protect the eye from the bright sunlight; this, however, merely facilitates observation of the bright planets, but not the stars.

The reason why stars are seen through the telescope in daylight is: not because they are seen from "the bottom of a tube," as many think, but because the refraction of light by the lens or its reflection in the mirrors detracts from the brilliancy of the part of the sky under observation, and al the same time enhances the brilliancy, of the stars (seen as points of light). We can see first-magnitude and even second-magnitude stars in daytime through a 7 cm. telescope. What has been said, however, does not hold true for either wells, mines, or chimneys.

The bright planets, say, Venus, Jupiter or Mars, in opposition, present a totally different picture. They shine far more brilliantly than the stars, and for this reason, given favorable conditions, can be seen in daylight (c.f. "The Planets in Daylight").

What Is Stellar Magnitude?

Even the layman with a hazy idea of astronomy knows that there are stars of the first magnitude and stars that are not of the first magnitude. These expressions are in general use. But he has scarcely heard of stars that are brighter than first-magnitude stars or what are known as zero or even minus magnitude stars. For him it would be out of all proportion to classify the brightest luminaries as minus magnitude stars, with the Sun a "minus 27th-magnitude star." Some even see in this a distortion of the negative number concept. And yet what we really have here is a striking illustration of the consistency of the theory of negative numbers.

First a few details about stellar classification by magnitude. I hardly need remind you that here the word "magnitude" means not the geometrical dimensions of the stars but their visual brilliancy. The ancients classified the brightest stars, those first seen in the evening sky, as stars of first magnitude. Then came stars of the second, third, fourth, fifth and, finally, sixth magnitude, which are roughly on the boundary line of unaided vision. This subjective classification of the stars by brilliancy did not satisfy the astronomers of later times. A more rigid basis for brilliancy classification was worked out. It was found that on the average the brightest stars (they are not all of equal brilliancy) are exactly 100 times brighter than the faintest stars on the borderline of unaided vision.

A scale of stellar brilliancy was established so that the ratio of brilliancy of stars of two neighboring magnitudes remains constant. Designating this "light intensity ratio" by n, we get:

2nd-mag. stars are fainter than 1st-magnitude stars by n times

3rd-mag. stars are fainter than 2nd-magnitude stars by n times

4th-mag. stars are fainter than 3rd-magnitude stars by n times, etc.

Comparing the brilliancy of stars of all other magnitudes with that of 1st-magnitude stars we get:

3rd-mag. stars are fainter than 1st-magnitude stars by n^2 times

4th-mag. stars are fainter than 1st-magnitude stars by n^3 times

5th-mag. stars are fainter than 1st-magnitude stars by n^4 times

6th-mag. stars are fainter than 1st-magnitude stars by n^5 times

Observation established that $n^5=100$. It is now easy (with the help of logarithms) to find the value of the light intensity ratio n:

$$n = 2.5^5$$

Thus, the stars of each successive magnitude are 2.5 times fainter than those of the foregoing magnitude.

Stellar Algebra

A few additional details about the group of brightest stars. We noted above that they are not of equal brilliancy; some are several times brighter than the average, others are fainter (their average brilliancy is a hundred times greater than that of stars on the borderline of unaided vision).

Let us now designate the brilliancy of stars 2.5 times brighter than the average first-magnitude star. What number precedes 1? Zero. Consequently, these stars are classified as "zero"-magnitude stars. But where should we place stars not 2.5 times, but only 1.5 times or twice brighter than those of first magnitude? They

---

-5- More accurately 2.512

come between 1 and 0, so their stellar magnitude is expressed by a positive decimal fraction, say, a star of "0.9 magnitude" or "0.6 magnitude," and so on. These stars are brighter than those of the first magnitude.

You will now realize why negative numbers had to be introduced in designating stellar brilliancy. Since there are stars of a light intensity *surpassing* the zero magnitude, their brilliancy, obviously, must be expressed in numbers on the other side of zero, i.e., by negative numbers. Hence such definitions of brilliancy as -1, -2, -1.6, -0.9, etc.

In astronomical practice "stellar magnitude" is ascertained by a device known as a photometer; the brilliancy of a luminary is compared with the brilliance of a definite star of known light intensity, or with an "artificial star" in the device itself.

The brightest star in the sky, Sirius, has a stellar magnitude of -1.6. Canopus (seen only in southern latitudes) has a stellar magnitude of -0.9. The brightest stars in the Northern Hemisphere are Vega (0.1), Capella and Arcturus (0.2), Rigel (0.3), Procyon (0.5) and Altair (0.9). (Remember that 0.5 magnitude stars are *brighter* than 0.9 magnitude stars, and so on.) Here is a list of the brightest stars and their stellar magnitude (the name of the constellation is given in brackets):

Sirfus (α Canis Majoris)	-1.6	Betelgeuse (α Orionis)	0.9
Canopus (α Carinae)	-0.9	Altair (α Aquilae)	0.9
α Centauri	0.1	α Crucis	1.1
Vega (α Lyrae)	0.1	Aldebaran (α Tauri)	1.1
Capella (α Aurigae)	0.2	Pollux (β Geminorum)	1.2
Arcturus (α Bootis)	0.2	Spica (α Virginis)	1.2
Rigel (β Orionis)	0.3	Antares (α Scorpii)	1.2
Procyon (α Car:is Minoris)	0.5	Fomulhaut (α Piscis Australis)	1.3

Achernar (α Eridani)	0.6	Deneb (α Cygni)	1.3
β Centauri	0.9	Regulus (α Leonis)	1.3

Glancing at this list we see that there are no stars at all exactly of first magnitude; from stars of 0.9-magnitude the list takes us to 1.1-, 1.2-magnitudes, etc., skipping the first magnitude. It follows, therefore, that the first-magnitude star does not exist, it is simply a conventional standard of brilliancy.

One should not take stellar magnitude classification as being determined by the physical properties of the stars. It derives from our eyesight and is the working out of the Weber-Fechner psycho-physiological law which is common to all the senses. Applied to vision this law reads: when luminosity changes in geometrical progression, light intensity sensation, changes in arithmetical progression. (Curiously enough, in measuring sound and noise intensity physicists rely on the principle applied in establishing stellar brilliancy; the reader will find this described in detail in my *Physics for Entertainment* and *Algebra for Entertainment*.)

Now that we have made the acquaintance of the astronomical scale of brilliancy, let us turn to some instructive calculations. Say we reckon how many third-magnitude stars taken together would shine with the brilliance of one first-magnitude star. We know that third- magnitude stars are fainter than first-magnitude stars 2.5^2 or 6.3 times; hence, to substitute for one first-magnitude star we would need 6.3 third-magnitude stars. Accordingly, we would need 15.8 fourth- magnitude stars and so on. The results of the calculations are given in the following table.[6]

To replace one first-magnitude star we need the following number of stars of other magnitudes:

2nd	2.5	7th	250

-6- The calculations are facilitated by the fact that the light intensity ratio logarithm is very simple, 0.4.

3rd	6.3	10th	4,000
4th	16	11th	10,000
5th	40	16th	1,000,000
6th	100		

At the 7th magnitude we are already over the borderline, in the world of stars inaccessible to the naked eye. Sixteenth-magnitude stars can be seen only through very powerful telescopes; to catch them with unaided vision, sensitivity of natural eyesight would have to be 10,000 times stronger. We would then see them as we now see the 6th-magnitude stars.

The above table could not, of course, provide for stars of the "pre- first" magnitude. Here are the calculations for some of them; 0.5-magnitude stars (Procyon) are $2.5^{0.5}$, or 1.5 times brighter than 1st-magnitude stars, -0.9-magnitude stars (Canopus) are $2.5^{1.9}$, or 5.8 times brighter, while -1.6-magnitude stars (Sirius) are $2.5^{2.6}$, or 11 times brighter.

Lastly, here is another interesting calculation: how many first-magnitude stars would be needed to replace the light shed by all the stars seen with the naked eye?

We take it for granted that there are 10 first-magnitude stars in one hemisphere. It has been observed that the number of stars in the category next in succession is roughly thrice the number of the previous category. In brilliancy they are 2.5 times fainter. Therefore the number we need would be equal to the sum of the members in the progression:

$$10 + (10 \times 3 \times 1/2.5) + (10 \times 3^2 \times 1/2.5^2) + \dots + (10 \times 3^5 \times 1/2.5^5) = 95$$

Hence, the, sum of the brilliance of all naked-eye stars in one hemisphere is roughly that of 100 first-magnitude stars or of one -4-magnitude star.

If we repeat the calculation so that it takes in not only the naked-eye stars, but also those accessible to modern telescopes, we shall find total brilliance to be equivalent to that of 1,100 first-magnitude stars or of one -6.6-magnitude star.

The Eye and the Telescope

Let us compare telescopic observation of stars with naked-eye observation.

We shall take the diameter of the human pupil in night observations to average 7 mm. A 5 cm telescope gates $(50/7)^2$ or roughly 50 times more light, and a 50 cm one 5,000 times more than the human pupil. Such is the number of times that the, telescope magnifies the brilliancy of the stars! (This applies solely to *stars*, not to planets, the discs of which can be seen. In calculating the brilliancy of the planetary image it is necessary to take into account the telescope's optical magnifying power.)

Knowing this, you will be able to reckon the diameter of the telescope lens needed to see stars of one or another magnitude; in addition, we must know what magnitude can be seen through a telescope with a definite lens. Suppose you know that a 64 cm telescope can take in stars up to the 15th magnitude inclusive. What lens diameter would be needed to see stars of the next, 16th, magnitude? In the ratio

$$x^2/64^2 = 2.5,$$

x is the unknown lens diameter. We find

$$x = 64\sqrt{2.5} \approx 100 \text{ cm}$$

Hence, we need a telescope with a lens diameter of one meter. Generally speaking, to increase the telescope's magnifying power to take in the next stellar magnitude, the lens diameter should be increased by $\sqrt{2.5}$ or 1.6 times.

Stellar Magnitude of Sun and Moon

Let us continue our algebraic excursion to take in celestial objects. Using the scale applied in estimating stellar brilliancy, we can, in addition to the fixed stars, find a place for the other luminaries -the planets, the Sun and the Moon. I shall dwell specially on planetary brilliancy later; meanwhile, let us talk about the stellar magnitude of the Sun and the Moon: The stellar magnitude of the Sun is expressed by the number -26.8, that of the full[7] Moon by -12.6. I take it that readers will have deduced from the foregoing pages why the two numbers are negative. What might be perplexing; however, is the seemingly inadequate difference between the stellar magnitude of the Sun and the Moon. You might say: the former "is only twice the latter."

Do not forget, however, that stellar magnitude is actually a kind of logarithm with a base of 2.5. And since it is impossible, when comparing numbers, to divide the logarithm of one by the other, similarly when comparing stellar magnitudes one cannot be divided by the other. The following calculation will show the result of correct comparison.

The Sun, with a stellar magnitude of -26.8, is $2.5^{27.8}$ times brighter than a first-magnitude star, while the Moon is but $2.5^{13.6}$ times brighter.

Hence, the Sun's brilliancy is $2.5^{27.8}/2.5^{13.6} \approx 2.5^{14.2}$ times that of the full moon. Computing this with the aid of a logarithm table we obtain the number 447,000. Consequently, the correct brilliancy ratio of the Sun and the Moon can be stated thus: In clear weather our diurnal luminary shines on the Earth with the power of 447,000 full Moons on a cloudless night.

If we take the quantity of heat reflected by the Moon to be proportional to the quantity of its light -and this probably approximates to truth- it can be presumed that it gives us 447,000

-7- In the first and last quarters the Moon's stellar magnitude is -9.

times less heat than the Sun. Astronomers know that each square centimeter on the borderline of the Earth's atmosphere receives from the Sun about two small calories of heat per minute. Consequently, the Moon supplies each square centimeter of the Earth with not more than the 225,000[th] fraction of a calorie per minute. In other words, it would heat one gram of water in one minute to the 225,000[th] fraction of a degree. We see, therefore, how groundless are the attempts to ascribe to moonlight any influence on the Earth's weather.[8]

The belief is widely current that clouds often melt under the rays of the full Moon. This is a crass delusion, the explanation being that nocturnal disappearance of clouds (for which there are other reasons) is noticed only in moonlight.

We shall now take leave of the Moon and calculate the number of times the Sun is brighter than Sirius, the brightest star in the sky. Proceeding in the same way as previously, we shall find the ratio of brilliancy to be:

$$2.5^{27.8}/2.5^{2.6} = 2.5^{25.2} = 10,000,000,000$$

i.e., the Sun is 10,000 million times brighter than Sirius.

Now for another interesting calculation. How many times brighter is the light shed by the full Moon than the combined light of all the stars in the firmament, i.e., all the naked-eye stars in one celestial hemisphere? We have already calculated that all the stars, from first to 6th magnitude inclusive, give as much light as 100 first-magnitude stars. Consequently, our task boils down to finding out the number of times the Moon is brighter than 100 first-magnitude stars. This is equal to

$$2.5^{13.6}/100 = 2,700$$

Hence, on a clear, moonless night the stars yield only $1/2700^{th}$ of the light shed by the full Moon, or 2,700 x 447,000,

-8- The moon gravitational influence on weather will be discussed at the end of this book (see "the Moon and the Weather")

i.e., 1,200,000,000 times less than the Sun on a cloudless day.

We might add that the star magnitude of a normal international "candle" one meter away is -14.2, signifying that at this distance it emits a light $2.5^{14.2-12.6}$, or four times brighter than a full Moon.

It will not be devoid of interest to note, moreover, that an aircraft beacon of 2,000 million candle-power would be seen, if placed at the Moon's distance from Earth, as a 4.5 magnitude star, or, in other words, could be seen with the naked eye.

True Brilliance of Stars and Sun

All our evaluations of brilliance so far have pertained solely to visual brilliance. The figures given express the brilliance of luminaries for their real distances. But we are, well aware that the stars are not all the same distance away. Hence, their visual brilliance tells us both true brilliance and the distance from us, or, to be more exact, of neither, until we separate the two. Meanwhile it is important to know what would be the comparative brilliance, or "luminosity," as they say, of the different stars were they all the same distance away.

Raising the question in this way astronomers introduce the concept of "absolute" stellar magnitude, that is, the magnitude of a star separated from us by a distance of 10 "parsecs." The parsec is a unit of length used to express the distances of stars; we shall speak of its origin later. For the time being we merely note that one parsec is equal to about 30,800,000,000,000 km. We can easily calculate absolute stellar magnitude if we know the distance to the star and take into account that brilliancy is in inverse proportion to the squared distance.[9]

-9- The following formula (the origin of which will be appreciated a little later when the reader becomes closer acquainted with the "parsec" and "parallax") can be used for the calculation: $2.5^{M} = 2.5^{m}(\pi/0.1)^2$ where M is the star's absolute magnitude, m its visual magnitude, and n the star's parallax in seconds. The sequence of operations in the calculation is as follows:

We shall acquaint the reader with the results of but two calculations, performed for Sirius and for the Sun. The absolute magnitude of Sirius is +1.3, that of the Sun +4.8. This means that at a distance of 308,000,000,000,000 km., Sirius would be seen as a 1.3-magnitude star and the Sun as a 4.8-magnitude star, being

$$2.5^{3.8}/2.5^{0.3} = 2.5^{3.5} = 25 \text{ times}$$

fainter than Sirius, though the Sun's visual brilliance is 10,000,000,000 times greater than that of Sirius.

We see, therefore, that the Sun is far from being the brightest star in the sky. But, let us beware of counting our Sun a pygmy among its fellow stars; for its luminosity is above the average. According to stellar statistics, average luminosity stars around the Sun, up to the distance of 10 parsecs away, are stars of the 9th absolute magnitude. Since the Sun's absolute magnitude is 4.8, this means that it is

$$2.5^{8}/2.5^{3.8} = 2.5^{4.2} = 50 \text{ times}$$

brighter than the average "neighboring" star.

Although 25 times fainter than Sirius in absolute values, the Sun, nevertheless, is 50 times brighter than the average stars around it.

Brightest of Known Stars

Greatest luminosity belongs to an 8th-magnitude star, invisible to the naked eye, in the Dorado constellation, designated by

---

$$2.5^{M} = 2.5^{m} \times 100\pi^{2}$$
$$M \log(2.5) = m \log(2.5) + 2 + 2 \log(\pi)$$
$$0.4 \, M = 0.4 \, m + 2 + 2 \log(\pi)$$
$$M = m + 5 + 5 \log(\pi)$$

For Sirius, for instance, m= -1.6 and π = 0.38″. Thus its absolute magnitude will be M= -1.6+5+5 log (0.38) =1.3

the Latin letter S. This constellation is in the Southern Hemisphere and cannot be seen in the temperate zone of the Northern Hemisphere. The star in question is a member of the neighboring stellar system of the Smaller Magellanic Cloud, which is roughly 12,000 times farther from us than Sirius. In view of the enormous distance it must have exceptional luminosity to be seen even as one of the 8th magnitude. If Sirius were cast so far away in space it would be seen as a 17th-magnitude star, or hardly seen at all even in the most powerful telescope.

What, then, is the luminosity of this remarkable star? Calculations give us this result: -9th magnitude. This means that in absolute terms it is roughly 400,000 times brighter than the Sun! If this most luminous of stars were at the same distance from us as Sirius it would be 9 magnitudes brighter, or about as brilliant as the Moon in its quarter phase! Certainly a star, which at Sirius' distance would shed such a volume of light on the Earth is entitled to be called the brightest of known stars.

Stellar Magnitude of Planets as Seen in Our Sky and in Alien Skies

Let us repeat the imaginary tour of the other planets that we made earlier in "Alien Skies" with a view to obtaining a more precise evaluation of the brilliance of the luminaries shining there. First, we shall indicate the planets' stellar magnitude at full brilliance in the Earth's sky. Here is the corresponding table:

In the Earth's Sky	
Venus	-4.3
Mars	-2.8
Jupiter	-2.5
Mercury	-1.2

Saturn	-0.4
Uranus	+5.7
Neptune	+7.6

This table shows us that Venus is nearly two stellar magnitudes, i.e., 2.5^2 or 6.25 times brighter than Jupiter, and $2.5^{2.7}$ or 13 times brighter than Sirius (Sirius' brilliance is of the -1.6 magnitude). It shows us also that the faint planet of Saturn is, nevertheless, brighter than all the fixed stars, save Sirius and Canopus. Here we find a pointer explaining why the planets Venus and Jupiter are sometimes accessible to unaided vision in daytime, while the stars are invisible to the naked eye in daylight.

Now for tables showing the brilliance of luminaries as seen in the skies of Venus, Mars and Jupiter, without additional explanations, since they would merely express in quantity what has already been said in "Alien Skies":

In Mars' Sky	
The Sun	26
Phobos	8
Deimos	3.7
Venus	3.2
Jupiter	-2.8
The Earth	-2.6
Mercury	-0.8
Saturn	0.6

In Venus' Sky	
The Sun	-27.5
The Earth	-6.6
Mercury	-2.7
Jupiter	-2.4

The Moon	-2.4
Saturn	-0.3

In Jupiter' Sky	
The Sun	-23
I Satellite	-7.7
II Satellite	-6.4
III Satellite	-5.6
IV Satellite	-3.3
V Satellite	-2.8
Saturn	-2
Venus	0.3

In evaluating the brilliance of the planets in the skies of their own satellites, pride of place should be assigned to the "full" Mars in the sky of Phobos (-22.5), then the "full" Jupiter in the sky of the V Satellite (-21) and the "full" Saturn in the sky of its satellite Mimas (-20); here Saturn is but five times fainter than the Sun!

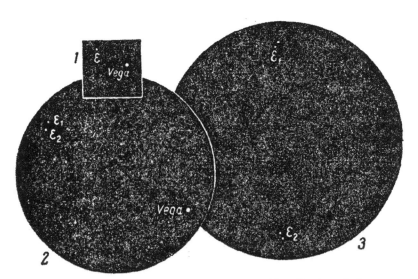

Fig. 73. The star ε Lyrae (near Vega) as seen (*1*) with the naked eye, (*2*) through a pair of binoculars, and (*3*) through a telescope.

Finally, another instructive table showing the brilliance of the planets as seen from one another. They are given here in the order of diminishing brilliancy.

Stellar Magnitude		Stellar Magnitude	
Venus from Mercury	-7.7	Mercury from Venus	-2.7
The Earth from Venus	-6.6	The Earth from Mars	-2.6
The Earth from Mercury	-5	Jupiter from the Earth	-2.5
Venus from the Earth	-4.3	Jupiter from Venus	-2.4
Venus from Mars	-3.2	Jupiter from Mercury	-2.2
Jupiter from Mars	-2.8	Saturn from Jupiter	-2
Mars from the Earth	-2.8		

The table shows that the brightest luminaries in the skies of the chief planets are Venus seen from Mercury, the Earth seen from Venus, and the Earth seen from Mercury.

Why Are the Stars Not Magnified in the Telescope?

People looking at the stars through a glass for the first time are surprised when the glass, which so noticeably magnifies the Moon and the planets, far from magnifying the stars, actually reduces them, turning them into bright points without a disc. This was noted in his day by Galileo, the first man to look at the sky with aided vision. Describing his early observations with the spy-glass which he had devised, he says:

"The difference in the shape of the planets and of the fixed stars when observed through la spy-glass is worthy of note. Whereas the planets are seen as small circles distinctly penciled, like minute coins, the fixed stars have no distinguishable outline.... The glass merely magnifies their brilliance, imparting to 5th- and 6th-magnitude stars the radiance of Sirius, the brightest of fixed stars."

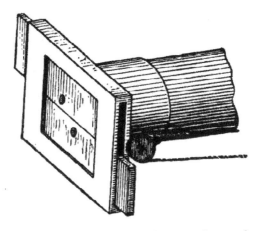

Fig. 74. A drawing explaining the working of the "interferometer," a device for
measuring the angular diameters of stars
(see text for details).

To explain this powerlessness of the telescope to magnify the
stars, we must remind you of a few facts concerning the physiol-
ogy and physics of eyesight. When your eyes follow a man walking away from you, his image on your retina becomes smaller
and smaller. At a certain distance his head and feet are so close
on the retina, that they no longer catch different elements (nerve
ends) but one and the same element. The human figure is seen
then as a dot, devoid of outline. This happens with most people
when the angle of vision diminishes to 1'. The purpose of the
telescope is to increase the angle of vision, or, in other words, to
elongate the image of each part of the observed object to embrace
several associated elements of the retina. We say of a telescope
that it "magnifies 100 times," when the angle of its vision is 100
times greater than the angle of unaided vision for the same distance. But when, even with this magnification, an object is seen
at an angle of less than 1', this tells us that the telescope is not
powerful enough to take in the object.

We can easily calculate that even the smallest detail seen on
the Moon through a telescope of 1000 x magnification would
have a diameter of 110 meters. For the Sun this would be 40 km.
But for the nearest star we would obtain the hugs diameter of
12,000,000 km.

The Sun's diameter is 8.5 times less. Thus, if transferred to the distance of the nearest star, the Sun would appear as a mere dot even in a 1000 x telescope. The nearest star must be about 600 times larger than the Sun to enable a powerful telescope to reveal its disc. At Sirius' distance, the star would be 5,000 times larger than the Sun. And since most stars are much farther away, and their average size not much greater than the Sun, they are seen as dots even through the most powerful telescopes.

"No star in the sky," Jeans says, "has a greater angular dimension than that of a pinhead 10 kilometers away and there is still no telescope which would show so small an object as a disc." On the contrary, the large celestial objects in our solar system, seen through the telescope, reveal a larger disc the greater the magnification. But, here, is we have already noted, the astronomer encounters another inconvenience; while the image increases, its brilliance diminishes (due to the scattering of the light beam over a greater surface), and the fainter brilliancy hinders detection of details. That is why when observing planets and comets particularly; the astronomer uses a telescope of moderate magnifying power.

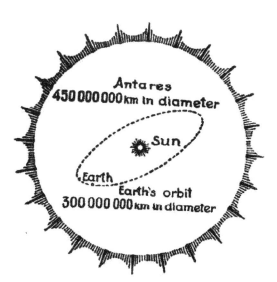

Fig. 75. The giant star Antares (α Scorpii) could encompass our Sun together with the Earth's orbit.

The reader may ask: Well, if the telescope does not enlarge the stars why use it at all?

After what has been said above there is hardly any need to dwell at length on this point. The telescope is powerless to magnify the apparent dimensions of the stars, but it intensifies their brilliance and, consequently, multiplies the number of stars accessible to vision.

Another achievement of the telescope is that it separates the stars which appear as one to the naked eye. Although it cannot enlarge the apparent diameter of the star, it enlarges the apparent distance between them. Thus, it reveals to us double, triple and still more complex stars in places where the naked eye sees but one (Figure 73). Stellar clusters which due to distance appear to the naked eye as hazy specks, or are not seen at all, break up into thousands of separate stars in the telescopic field of vision.

A third service rendered by the telescope in stellar study is that it enables astronomers to measure angles with amazing accuracy; on photographs obtained with the help of big modern telescopes astronomers measure angles of o."o1. At this angle we can see a farthing 300 km. away, or a human hair 100 meters away!

How Were Stellar Diameters Measured?

As we have just explained the diameters of the fixed stars cannot be seen even with the most powerful telescope. Until recently conjectures about stellar dimensions were simply guesswork. It was presumed that each star approximated to our Sun in size, but corroboration was lacking. And since telescopes more powerful than those in use were needed to measure the stellar diameters, it seemed that the problem of ascertaining the true diameters of the stars was insoluble.

That was how matters stood until 1920 when new methods and investigation instruments enabled astronomers to tackle the job.

For this latest achievement astronomy is indebted to its loyal ally, physics, which has rendered it more than one inestimable service.

We shall now describe the gist of the method based on light interference.

To elucidate the principle on which this method of measurement is based we shall conduct an experiment calling for a few simple tools, including a small 30 x telescope and a bright light source located 10 or 15 meters away and fenced off by a screen with the narrowest of vertical slits -a few tenths of a millimeter. We now cover the lens with an opaque lid in which there are two round apertures about 3 mm in diameter and 15 mm apart horizontally symmetrical to the center of the lens (Figure 74). When the lid is removed, the slit is seen in the telescope as a narrow band with far fainter side strips. But when we look at the slit with the lid on, we see vertical dark gaps on the central bright band -the consequence of the interaction (interference) of the two beams of light passing through the two apertures in the lid; they vanish when we cover one of the apertures.

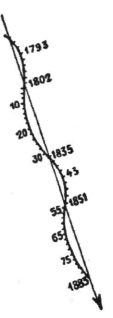

Fig. 76. The path followed by Sirius among the stars between 1793 and 1922.

If the apertures are so placed that the distance between them can be changed, we find that the farther apart they are the fainter the dark gaps become, until at last they vanish altogether. Knowing the distance separating the /apertures, we can ascertain the angular width of the slit, or in other words, the angle at which this width is seen by the observer. And if, in addition, we know

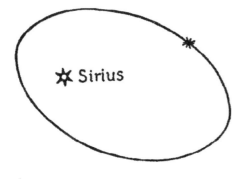

Fig. 77. The orbit of Sirius' satellite with respect to Sirius itself. (The reason why Sirius is not at the focus of the apparent ellipse here is because the true ellipse has been distorted in projection; here we see it at an angle.)

the distance to the slit itself, we shall be able to calculate its real width. But if instead of a slit we have a tiny round hole, even then the method of ascertaining the width of the "circular slit" (i.e., the diameter of the circle) remains the same, all that is needed is to multiply the angle obtained by 1.22.

The same method is used in measuring the diameter of stars, but since the angular diameter of the latter is so tiny, a very powerful telescope is needed.

In addition to this instrument, which is known as the "interferometer," there is another more roundabout way of ascertaining the true diameter of the stars, based on investigation of stellar spectra.

The astronomer deduces from the star's spectrum its temperature and hence is able to calculate the quantity of radiance emanating from 1 sq. cm of its surface. Moreover, if he knows the distance to the star and its visual brilliance, he can ascertain the quantity radiated by the entire surface. The ratio of this second quantity to the first gives us the area of the star's surface and, hence, its diameter. It has been established, for example, that the diameter of Capella is

Fig. 78. Sirius' satellite consists of a substance 60,000 times denser than water. A few cubic centimetres of it would weigh the same as thirty men.

16 times the Sun's diameter; Betelgeuse's diameter is 350 times bigger; Sirius's diameter and Vega's diameter are two and a half times bigger while the diameter of Sirius' satellite is 0.02 times the Sun's diameter.

Giants of the Stellar World

The results obtained when stellar diameters were measured were truly startling. Astronomers had no idea that the universe contained such huge stars. The first star to have its true dimensions successfully recorded (1920) was the bright α Orionis, bearing the Arabic name Betelgeuse. Its diameter proved to be greater than the diameter of Mars' orbit! Another of the giants is Antares, the brightest star in the Scorpius constellation; its diameter is about 1.5 times that of the Earth's orbit (Figure 75). Among the stellar giants discovered so far mention should be made of the Cetus constellation, whose diameter is 400 times greater than that of the Sun.

A few words about the physical structure of these giants. Calculations show that these stars, monster dimensions notwithstanding, have disproportionately small amounts of matter. They are only a few times heavier than the Sun, and since Betelgeuse, for instance, is 40,000,000 times larger than the Sun, its density, consequently, must be negligible. And if the Sun's matter averages the density of water, then, the matter of the giant stars must be in the nature of rarified air. As one astronomer put it, these stars "resemble huge balloons with meager densities, much smaller than that of air."

An Unexpected Result

In connection with the foregoing it would be interesting to know ill how much space all the stars would fill in the sky if their visual images were juxtaposed.

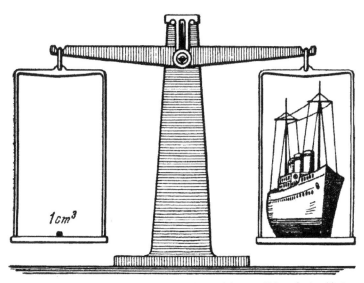

Fig. 79. One cubic centimetre of atomic nuclei, even if loosely jumbled together, would have the weight of an Atlantic liner. If packed tightly it would weigh 10 million tons!

We know that the aggregate brilliance of all the stars .accessible to telescopic observation is equal to that of a -6.6 magnitude star (see above). The luminosity of such a star is 20 stellar magnitudes fainter 4 than that of the Sun, in other words it gives off 100 million times less light. Assuming the Sun, in surface temperature, to be an average star we can take the apparent surface of our imaginary star to be the indicated number of times less than the apparent surface of the Sun. Since diameters of circles are proportional to the square roots of their areas, the apparent diameter of our star would be 10,000 times less than the apparent diameter of the Sun, i.e., 30':10,000=0.2".

The result is astounding: the total apparent area of all the stars fills as much space as a disc with a 0.2" angular diameter. The sky contains 11,253 sq. degrees; from this we easily deduce that the stars accessible to the telescope cover but 1/20,000 millionth of the entire sky!

The Heaviest Substance

Among the wonders concealed in the bosom of the universe a prominent place is likely to be held for all time by the miniature star located near Sirius. This star consists of a substance 60,000 times heavier than water! When you take a glass of mercury in your hand you are surprised by its weight -some 3 kg. But I wonder what you would say about a glassful of a substance weighing 12 tons and needing a railway car to carry it? It sounds absurd, doesn't it? Yet this is one of astronomy's latest discoveries.

The discovery, incidentally, has a long and very instructive history. It was observed long ago that the brilliant Sirius moved among its fellows not along the straight line that most stars follow, but along a strange, tortuous path (Figure 76). To explain this feature of its motion, the famous astronomer Bessel conjectured that Sirius had an attendant satellite whose gravitation "perturbed" its motion. That was in 1844, two years before the "pen-nib" discovery of Neptune. In 1862, when Bessel was no longer living, his surmise was fully corroborated, when Sirius' suspected satellite was spotted in the telescope.

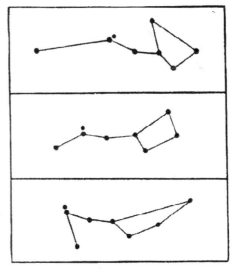

This satellite, the so-called "Sirius B," revolves around its primary in a period of 49 years at a distance 20 times farther than the Earth from the Sun (roughly the distance of Uranus from the Sun) (Figure 77). Although this is a faint 8th- or 9th-magnitude star, it has an impressive mass -almost 0.8 the mass of the Sun. If the Sun was as distant as among the stars Sirius is, its luminosity would be that of a 1.8-magnitude star. Consequently, if the area of Sirius' satellite

Fig. 80. As the aeons roll on, the contours of the constellations slowly change. The middle drawing depicts the "Dipper" of the Great Bear as it is now, the upper—as it was 100,000 years ago, while the lower drawing shows what it will be like 100,000 years hence.

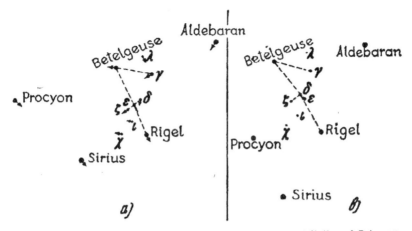

Fig. 81. The directions in which the bright stars of the constellation of Orion are moving (*a*), and how these motions will have changed the shape of the constellation in 50,000 years from now (*b*).

were proportionately the same number of times less the area of the Sun as its mass is, it would, given the same temperature, have the luminosity of roughly a second-magnitude star, not of an 8th- or 9th-magnitude star. Astronomers first attempted to explain the faint brilliance by presuming a low surface temperature; it was thought to be cold sun covered with a hard crust.

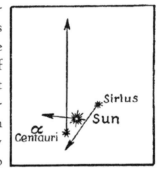

Fig. 82. The direction of the three neighbouring stars, the Sun, α Centauri and Sirius.

This conjecture, however, proved wrong. Thirty years ago it was found that Sirius' unassuming satellite was by no means a fading star, but, on the contrary, belonged to the category with a high surface temperature, far higher than our Sun. This changed the picture altogether, and gave grounds for ascribing the faintness solely to its small surface dimensions. Calculations have shown that it emits 360 times less light than the Sun; consequently its area must be at least 360 times less than that of the Sun, and its radius $\sqrt{360}$, or 19 times less. So we conclude that in volume Sirius' satellite is less than a $1/6,800^{th}$ of the Sun; in its mass, however, it is nearly 0.8 that of our diurnal luminary. This alone points to the heavy density of the substance of this star. More accurate calculations show this

Fig. 83. A scale of stellar motions. Two croquet balls, one in Leningrad, the other in Tomsk, move towards each other with the speed of 1 km. a century—such in miniature is the approach towards each other of two stars. This shows how remote is the possibility of stars colliding.

-planet to have a diameter of only 40,000 km., and, consequently, a density in the realm of the monster previously mentioned -60,000 times the density of water (Figure 78).

"Prick up your ears, physicists: your domain is about to be invaded." Kepler's words, true, they refer to another matter, came to mind. Indeed no physicist had ever imagined anything of the like. In ordinary conditions such enormous density is absolutely unthinkable: the space between normal atoms in solid bodies is too small to allow for any noticeable compression. It is a different matter, however, with "mutilated" atoms that have parted with the electrons circling around the nuclei. Loss of electrons reduces atomic diameters several thousand fold, with hardly any effect on the masses; the shorn nucleus is about the same number of times smaller than the normal atom as a fly is smaller than a big building. Subjected to the terrific pressures in the depths of the stellar sphere, these diminished atom-nuclei can come thousands of times closer to each other than normal atoms and produce a substance of unheard-of density such as that disclosed on Sirius' satellite. What is more, this density has been surpassed by the so-called van-Maanen star, which though of the 12th magnitude and no larger than the Earth, consists of a substance possessing a density 400,000 times that of water!

But this is still not the extreme limit. Theoretically, we can presume far denser substances to exist. The diameter of the atomic nucleus is not more than 1/10,000[th] the diameter of the atom and, consequently, its volume is not more than 1/10^{12} that of the atom. One cubic meter of metal contains only 1/1000 mm³ of nuclei, and the metal's entire mass is concentrated in this minute volume. Accordingly, 1 cu. cm. of nuclei should weigh roughly 10,000,000 tons (Figure 79).

After what has been said there will be nothing incredible about the discovery of a star with an average density 500 times greater than that of the afore-mentioned Sirius B. I have in mind the tiny 13th-magnitude star discovered in the Cassiopeia constellation at the end of 1935. Though 1/8th the size of the Earth and no bigger than Mars, this star has a mass nearly three times (2.8 times to be exact) that of the Sun. In ordinary units its mean density is expressed by the number 36,000 000 gr/cm^3. This means 1 cm^3 of this substance would weigh 36 tons on Earth! Consequently, it has a density nearly two million times that of gold.[10] In Chapter Five we shall discuss how much a cubic centimeter of this substance would weigh on the star itself.

Only a few years ago scientists would, of course, have scouted the idea of a substance with a density millions of times that of platinum.

But it is quite on the cards that the fathomless depths of the universe still conceal not a few wonders of this kind.

Why Are Stars Called Fixed Stars?

When the ancients gave the stars this name they wished to emphasize that in contrast to the planets the stars are fixed in the sky. Naturally, they participate in the daily motion of the heavens around the Earth, but this apparent motion does not disturb their related position. The planets, on the other hand, constantly change place in relation to the stars, wander in their midst and for this reason became known in antiquity as "wandering stars" (the literal meaning of the word "planet").

-10- The density in the centre of this star must be unbelievably great: about 1,000 million grams in 4 cu. cm.

We moderns know that this picture of the stellar world as an assemblage of immobile suns is absolutely wrong. All the stars,[11] including the Sun, move relative to each other with a mean velocity of 30 km. per sec., the speed of our planet in its orbital flight. Hence, the stars are no less mobile than the planets. Indeed in the world of stars we sometimes meet velocities of an order never found in the family of planets: astronomers know of "flying" stars which travel with respect to our Sun at the terrific speed of 250-300 km per sec.

But if all the visible stars move chaotically at breakneck speed, covering thousands of millions of kilometers annually, why do we not see this frantic flight? Why do the stellar heavens present a picture of majestic immobility?

The reason is simple: it is the incredible distance to the stars. Have you ever watched from a high point a train moving in the distance near the horizon? Did you not have the impression that the express was crawling along like a turtle? A breath-taking speed for a nearby observer seems a snail's pace when watched from a distance. It is the same with stellar motion; the only difference here is that the relative distance of the observer from the moving body is infinitely greater. Even the brightest stars, which on the average are much nearer to us than the others, being (according to Kapteyn) a mere 800 million kilometers away, shift within a year some 1,000 million kilometers or 800 thousand times less their distance away from us. To see this shift we on Earth would have to look at it from an angle of 0.25", a value hardly to be caught even by the most delicate of astronomical instruments. To the naked eye it would be utterly invisible, even if it continued for centuries. Only painstaking measurement with the most delicate instruments disclosed the motions of numerous stars (Figures 80, 81, 82).

Thus, when speaking of naked-eye observation, the "fixed stars," despite their incredibly rapid motion, are fully entitled to the name. Readers will gather from what has been said how utterly remote is the probability of stars colliding, notwithstanding

-11- Meaning the stars composing "our" stellar island—the Milky Way.

their dizzy speed (Figure 83)

Units of Stellar Distances

Our biggest units of length, the kilometer, the nautical mile (1,852 m) and the geographical mile (four nautical miles), while adequate for measuring length on the globe, are certainly not the thing for celestial distances. They would be about as inconvenient for measuring celestial distances as, say, millimeters for measuring the length of a railway line. The distance in kilometers from Jupiter to the Sun is 780,000,000. If we were to measure the Leningrad-Moscow Railway in millimeters we would get the number of 640,000,000.

To avoid a long row of naughts, astronomers use far bigger units of length. In measuring length within the limits of the solar system, they take as the unit the mean distance from the Earth to the Sun-149,500,000 km. This is the so-called "astronomical unit." In these units the distance between the Sun and Jupiter is 5.2, between the Sun and Saturn 9.54 and between the Sun and Mercury 0.387.

α Centauri

But when it comes to measuring distances between our Sun and other suns, the unit just mentioned is also too small. For instance, the distance to the nearest star (the so-called Proxima in the Centaurus constellation,[12] a reddish 11th-magnitude star) can be expressed in 260,000 such units.

But this is only the nearest star; the others are much farther away. Bigger units have greatly simplified the memorizing of these figures and their use. In astronomy we have the following giant units of length: the "light year" and the "parsec" which is now superseding the former.

Proxima Centauri

Fig. 84. The Sun's closest stellar system— α | Centauri, including A,B and Proxima Centauri.

-12- The bright star a Centauri is almost next to it.

The light year is the path covered in space by a ray of light within one year. We can get an idea of the size of this unit when we recall that it takes but eight minutes for sunlight to reach the Earth. A "light year," then, is greater than the radius of the Earth's orbit the number of times that the year is greater than eight minutes. In kilometers this unit will be 9,460,000,000,000, which means that the light year is approximately equal to 9½ billion km.

The *parsec*, the other unit favored by astronomers in measuring stellar distances, has a more complicated origin. The parsec is the distance which must be covered to see the semi-diameter of the Earth's orbit at the angle of one angular second. In astronomy the angle from which the semi-diameter of the Earth's orbit is seen from a star is known as the "annual parallax" of this star. The word "parsec" derives from combining the word "parallax" with the word "second." The parallax of a Centauri, the above-mentioned star, is 0.76 seconds; we can easily visualize that the distance to this star is 1.31 parsecs. It requires no great effort to work out that 1 *parsec* is equal to 206,265 times the distance from the Earth to the Sun. The relation of the parsec to other units of length is:

1 parsec = 3.26 light years = 30,800,000,000,000 km.

The distances to some of the bright stars, expressed in parsecs and in light years, is as follows:

	Parsec	Light year
α Centauri	1.31	4.3
Sirius	2.67	8.7
Procyon	3.39	11.0
Altair	4.67	15.2

These are the comparatively nearest stars. You will appreciate just how "near" they are, if you bear in mind that to express these distances in kilometers you should multiply each figure in the first column by 30 billion (a billion being a million millions). But not even the light year and parsec are the biggest units used in stellar astronomy. When astronomers began to measure the distances and dimensions of star clusters, that is, of "universes" within the universe, consisting of millions and millions of stars, they needed a still larger unit. They formed this from the parsec in the same way that we form the kilometer from the meter, thus obtaining the "kiloparsec" equal to 1,000 parsecs or 30,800 billion km. In these units the diameter of the Milky Way can be expressed, for instance, by the number 30, and the distance from us to the Andromeda nebula by a number around 300.

But soon the kiloparsec was found inadequate and astronomers were obliged to introduce the *"megaparsec,"* which equals *one million* parsecs.

Here is a table of stellar units of length:

1 megaparsec = 1 million parsecs
1 kiloparsec = 1 thousand parsecs
1 parsec = 206,265 astronomical units
1 astronomical unit = 149,500,000 km

It is impossible to visualize the megaparsec. If we were to reduce the kilometer to the thickness of a human hair (0.05 mm), the mega- parsec even then would be too much for the human imagination, it would be equal to 1,500,000,000 km, 10 times the distance from the Earth to the Sun.

Here, incidentally, is a comparison that will enable the reader to grasp the immensity of the megaparsec. The finest strand of a spider-web, stretched from Moscow to Leningrad, would weigh about 10 gr., and from the Earth to the Moon not more than 6 kg. The same strand stretched to the Sun would weigh 2.3 tons. But if it could be stretched the length of one megaparsec it would weigh 500,000,000,000 tons!

The Nearest Stellar Systems

A fairly long time ago, about a century back, it was established that the nearest stellar system was the first-magnitude binary star of the southern constellation of Centaurus. Recent years have added interesting details to our knowledge of this stellar system. A small 11th-magnitude star was found near α Centauri, thus comprising with the two stars of α Centauri a triple-star system. The fact that the third star is physically a member of the α Centauri system, despite their being more than 2° apart, is confirmed by the identity of motion; all-three stars travel with the same speed in the same direction. The most remarkable feature of the third member of this family is that it is nearer, to us than the other two and for this reason must be .regarded as the nearest of any of the stars whose distances have been ascertained so far. Hence the name "Nearest," in Latin "Proxima." It is 3,960 astronomical units nearer to us than the α-Centauri stars (α Centauri A and α Centauri B). Here are their parallaxes:

α Centauri (A and B): 0.751
Proxima Centauri: 0.762

Since A and B stars are only 34 astronomical units part, the entire system has rather strange appearance depicted on Figure 84.

A and B are slightly farther apart than Uranus and the Sun. Proxima, however, is 59 "light days" distant from them. These stars slowly change location: while the period of revolution of the A and B stars about their common centre of gravity is 79 years. Proxima makes one revolution in more than 100,000 years, so we need have no fear of it shortly ceasing to be our nearest star, of yielding place to another member of α Centauri family.

What do we know about the physical properties of the stars of this system? In brilliance, mass and diameter, α Centauri A is just a little ahead of the Sun (Figure 85), whereas α Centauri B, with a mass slightly smaller than the Sun, is 1/5th bigger in diameter. In brilliance, however, it is but a third of the Sun. Accordingly, its surface temperature is lower, being 4,400°C, while the Sun's is 6,000°C.

Proxima is still "colder"; its surface temperature is 3,000°C and it has a reddish color. It is 14 times smaller than the Sun in diameter and, though hundreds of times greater in mass, is smaller in size than Jupiter and Saturn. From α Centauri A we would see its partner B as being roughly the same size as the Sun seen in Uranus' skies. We would also see Proxima, but as a tiny, faint star, since it is 250 times farther than Pluto from the Sun and 1,000 times farther than Saturn from the Sun.

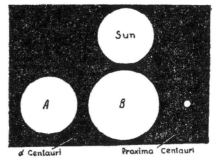

Fig. 85. The dimensions of the α Centauri stars and the Sun compared.

The sun's next nearest neighbor after the triple α Centauri star is the tiny 9.7-magnitude star in the Ophiuchus constellation known as the "Flying Star."

It gained its name from its exceedingly fast apparent motion. It is 1½ times farther away from us than is the α Centauri system, but is our nearest neighbor in the Northern Hemisphere of the sky. Its flight, at a tangent to the Sun's motion, is so sweeping that in a matter of under ten millenniums it will be twice as near, and will then be nearer than the triple α Centauri star.

The Scale of the Universe

Let us return now to that diminutive model of the solar system that we visualized according to the instructions furnished in the chapter on the planets; we shall augment it to include the world of stars. What do we get?

You will recall that on our model the Sun was a ball 10 cm.

in diameter, while the entire planetary system was represented by a circle 800 m. in diameter. At what distance from the Sun should we place the stars, providing we adhere to the same scale? We can easily calculate that we must place Proxima Centauri, our nearest star, at a distance of 2,700 km., Sirius, 5,500 km. and Altair 9,700 km. away. Even on our model these "nearest" stars would be crowded in Europe. For the more remote stars we shall take a unit bigger than the kilometre-1,000 km., or the Megameter (Mm). There are but 40 of these units in the Earth's circumference and 380 between the Earth and the Moon. On our model Vega would be 17 Mm away, Arcturus 23 Mm, Capella 28 Mm, Regulus 53 Mm and Deneb (α Cygni) more than 350 Mm away.

Reducing this last number of 350 Mm to kilometers we get 350,000 or a little less than the distance to the Moon. So our diminutive model, on which the Earth is a pinhead and the Sun a croquet ball, itself acquires cosmic dimensions!

But we have still to complete our model. The outlying stars of the Milky Way would be 30,000 Mm distant on our model, or nearly 100 times more than the distance to the Moon. But even the Milky Way is not the whole of the universe. Far beyond it are other stellar clusters as, for instance, the naked-eye system in the Andromeda constellation, or the equally visible Magellanic Clouds. On our model we shall designate the Smaller Magellanic Cloud by an object 4,000 Mm in diameter, and the Greater Magellanic Cloud by an object with a diameter of 5,500 Mm, placing the two 70,000 Mm away from the model of the Milky Way. We would have to lend the model of the Andromeda nebula a diameter of 60,000 Mm, placing it at a distance of 500,000 Mm from the model of the Milky Way, almost the actual distance from Jupiter to the Sun!

The remotest celestial objects for modern astronomy are the stellar nebulae -clusters of stars located far beyond the confines of the Milky Way and more than 1,000,000,000 light years from the Sun. The reader, if he wishes, can have a go at designating this distance on our model. If he succeeds he will have an idea of the dimensions of the part of the universe within range of the optical resources of modern astronomy.

The reader will also find a number of pertinent comparisons in my book *Do You Know Your Physics.*

CHAPTER FIVE

GRAVITATION

Shooting Vertically

Where would a cannon ball, discharged vertically from a gun mounted on the equator, fall (Figure 86)? This problem was discussed in a magazine some 20 years ago; it visualized that a cannon ball discharged with the starting velocity of 8,000 m per second, would, in the space of 70 minutes, reach a height of 6,400 km (the Earth's radius). This is what the magazine said:

Fig. 86. The problem of the vertically discharged cannon ball.

"If a cannon ball were discharged vertically from the equator, it would upon emerging from the barrel acquire in addition the east-directed circular velocity of the points at the equator (465 m/sec). That is the velocity with which the cannon ball would travel parallel to the equator. The point 6,400 km directly above the barrel at the moment of discharge would have moved along the circle of the double radius twice the velocity. It would therefore be ahead of the cannon ball in its eastward flight. When the latter reaches culminating height, it would not be directly above the point of discharge, but some distance behind, in the west. The process is repeated when the cannon ball returns to Earth. After 70 minutes' ascent and descent the cannon ball would fall at a point about 4,000 km to the west. That is where we should anticipate it. For the cannon ball to return directly into the barrel it should be discharged not vertically, but at a slightly tilted angle, 50 in our case."

Flammarion in his *Astronomy* solves the same problem in a totally different way:

"A cannon ball discharged straight into the air, to the zenith, will return to the gun barrel, even though during its ascent and descent the gun has moved eastward together with the Earth. The reason is self-evident. The ascending cannon ball loses nothing of the velocity imparted by the Earth's motion. The two impetuses it receives are not counteracting; it may ascend a kilometer and simultaneously travel, say, 6 km. east. Its motion in space will follow the diagonal of a parallelogram, of which one side is 1 km and the other 6 km. Its downward flight under gravitation will follow another diagonal (or rather as curve since the fall is accelerated) and the cannon ball will drop right into the barrel."

Flammarion adds: "However, the experiment would not be an easy one, because well calibrated guns are rare, and there would be difficulties in training them vertically. Mercein and Petit tried the experiment in the 17th century but failed even to recover the discharged ball. The title page of Varignon's *More Thoughts about Gravitation* (1690) featured a pertinent drawing (we reproduce it as an introduction to this chapter). It depicts two observers, a monk and a soldier, standing alongside a gun trained towards the zenith and looking upwards as if following the cannon ball's flight. The engraving carries the inscription in French: "Will it return?" The monk is Mercein and the soldier, Petit. They repeated the rather dangerous experiment several times, but, being bad marksmen they failed to make the cannon ball hit them on the head, so they came to the conclusion that it had remained somewhere in the air. Varignon exclaims in astonishment: 'A cannon ball hanging over our heads! What an astonishing thing' When this experiment was repeated in Strassburg, the cannon ball was found several hundred meters away from the gun. Apparently it had not been trained strictly vertically."

As we see, the two solutions clashed sharply. One claimed the ball would fall far to the west of the gun, the other that it was bound to return to the point of discharge. Which is correct?

Strictly speaking, both are wrong, though Flammarion came much closer to the truth. The cannon ball would fall west of the gun, but not at the distance the first claimed, nor, as the second thought, would it return to the barrel.

The solution to our regret is out of bounds to elementary mathematics.[1] So we shall confine ourselves to the final result.

Designating the cannon ball's starting velocity by v, the angular velocity of the Earth's rotation by ω and the gravity acceleration by g, we obtain for the distance x, the point where it falls west of the gun, the expression $x = (4/3)\ \omega\ v^3/\ g^2$ for the equator, and $x = (4/3)\ \omega\ v^3/\ g^2\ cos\ (\varphi)$ for the latitude φ.

In the problem set by the first author we know that $\omega = 2\pi/86,164$, v =8,000 m/sec and g=9.8 m/sec^2.

Accordingly, we find x =520 km; hence, the cannon ball will fall 520 km west of the gun (and not 4,000 km. away as the first author thought).

What answer does the formula furnish in Flammarion's case? The gun was discharged not on the equator, but near Paris, on the 48th parallel. We assume that the initial velocity of the cannon ball fired from the ancient gun was 300 m/sec. Knowing that $\omega = 2\pi/86,164$, v =300 m/sec, g=9.8 m/sec^2 and φ =48°, we find x = 18 m, the cannon ball will fall not back into the barrel as the French astronomer presumed, but 18 m. west of the gun. We have, of course, ignored the possible deflection caused by air currents which might tangibly alter the result.

Weight at High Altitudes

In the foregoing calculations, incidentally, cognizance is taken of a consideration which so far has not been mentioned. I have in mind the decline in gravitation the farther away we are from the Earth. Weight is simply a manifestation of universal gravitation, and with the increase in the distance between two bodies

-1- A special and detailed calculation would be needed. It was worked out by specialists at my request, but the details would take too much space here.

the mutual attraction rapidly decreases. According to the Newtonian law, gravitation is inversely proportional to the squared distance; the distance here is from the centre of the Earth, for the Earth attracts all bodies, as if its entire mass was concentrated in the centre. For this reason gravitation at a height of 6,400 km., at a point twice the Earth's radius away from its centre is but a fourth of the Earth's surface gravitation.

In the case of an artillery shell discharged upwards, this is manifested in the shell travelling higher than it would if gravitation did not diminish with altitude. We took it for granted that a cannon ball discharged vertically with an initial velocity of 8,000 m/sec. would reach a height of 6,400 km. But in reckoning the ceiling according to the universally known formula, disregarding the decline in gravitation as latitude increases, we would obtain a height half of what it really is. Here is the calculation. Textbooks on physics and mechanics give the following formula for computing the height h reached by a body discharged vertically with the velocity v and the constant acceleration of gravity g:

$$h = v^2/2g$$

If v = 8,000 m/sec. and g = 9.8 m/sec², h will equal to 3,265 km.

This is almost half the height indicated above. As already noted, the difference is due to our disregard in applying the textbook formula for the decline in gravitation caused by altitude. It is quite plain that if the Earth's attraction for the cannon ball diminishes, then with the given velocity it is bound to rise higher.

We must not conclude, however, that text-book formulas for the height of a body discharged vertically are wrong. They are correct within the limits for which they are intended, becoming unreliable only when used outside the indicated boundary. These formulas are designed for low altitudes, where decrease in gravitation is so infinitesimal that it can be discounted. For instance, in the case of a cannon ball discharged upwards with an initial velocity of 300 m/sec., the decrease in gravitation is of little import.

The interesting question is this: Is decline in gravitation felt at the altitudes reached by modern aircraft? Is loss of weight noticeable at these altitudes? In 1936 the pilot Vladimir Kokkinaki flew with varying payloads to high altitudes -half a ton to an altitude of 11,458 meters, 1 ton to 12,100 m. and 2 tons to 11, 295 m. The question is: Did these loads retain their original weight at the indicated altitudes, or did they lose tangibly? At first glance it might seem that an ascent 10 odd kilometers up would not appreciably decrease weight on such a vast planet as our Earth. On its surface the freight was 6,400 km. away from the centre of the planet. Its elevation by 12 km. increased the distance to 6,412 km. The addition seems too tiny to make any loss in weight felt. Calculations, however, showed the reverse. Loss of weight was quite tangible.

Let us make the, calculation for one particular case, say, for Kokkinaki's ascent with 2,000 kg to a height of 11,295 m. At this altitude the airplane was $(6,411.3/6,400)$ times farther from the centre of the Earth than at take-off.

Here gravitation is $(6,411.3/6,400)^2$, i.e. $(1+ 11.3/6,400)^2$ times less.

Hence at this height the freight would weight 2,000: $(1+ 11.3/6,400)^2$ kg.

Performing this operation (for which a rough and ready reckoning is quite suitable[2]) we find that at ceiling the 2,000 kg freight weighed only 1,993 kg, 7 kg less, quite a tangible loss. At this altitude a kilogram would indicate only 996.5 gr. on a spring balance, losing 3.5 gr.

The crew of out stratosphere balloon who reached an altitude of 22 km must have lost even more weight, namely, 7 grs per kg.

-2- We used the approximated equalities $(1+a)^2= 1+2a$, and $1/(1+a)=1-a$, where a is a minute quantity, where a is minute quantity. Therefore 2,000 : $(1+ 11.3/6,400)^2$ = 2,000: $(1+ 11.3/3,200)$ = 2,000 − 11.3/1.6 = 2,000 − 7

With Compasses Along Planetary Paths

Of the three laws of planetary motion which the painstaking genius of Kepler wrested from nature, the first is probably the least comprehensible to many. This law states that the planets move along elliptical paths. But why elliptical? It would seem that since the Sun exerts the same force in every direction and, moreover, a force that diminishes equally all round with distance, the planets should move about the Sun in circles, and not along extended locked paths, in which the Sun, it should be added, does not hold the central position. Mathematics explains the perplexities. But since not all amateur astronomers are at home with the calculus I shall help the reader to an understanding of the correctness of Kepler's laws.

With a pair of compasses, a scale ruler and a large sheet of paper let us chart the planetary paths and obtain visual proof that their outlines conform to Kepler's laws.

Gravitation guides planetary motion. Let us probe into this. The circle to the right on Figure 87 depicts an imaginary Sun; on the left we have an imaginary planet. Suppose the distance between them is 1,000,000 km.; on our drawing this is 5 cm, a scale of 200,000 km to 1 cm.

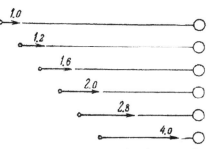

Fig. 87. The nearer a planet approaches the Sun the greater the gravitational pull of the latter.

The 0.5 cm. arrow depicts the force with which the Sun pulls at our planet (Figure 87). Suppose now that due to this attraction our planet has drawn closer to the Sun and is now only 900,000 km. away, or 4.5 cm on our drawing. The Sun's pull on the planet should have increased according to the law of gravitation by $(10/9)^2$, or 1.2 times. If, at the outset, we took an arrow of 1 unit to depict the force of gravitation, we must now use 1.2 units for our arrow. When the distance drops to 800,000 km -4 cm. on our drawing- the force of attraction increases $(5/4)^2$ or 1.6 times, and is depicted by an arrow of 1.6 units. In approaching a distance of respectively 700, 600 and 500 thousand kilometers from the Sun, the force of attraction is depicted by arrows respectively 2, 2.8 and 4 units of length.

One can imagine the same arrows indicating not only the attracting force but also the shift in position made by the body under the influence of this force within a unit of time (in this case the shift is proportional to the acceleration and consequently, to the force). We shall use this drawing, as a ready scale for planetary shifts, later on, in other drawings.

Let us now chart the path for a planet revolving around the Sun. Suppose that at a certain moment, a planet of the same mass as that mentioned above, moving in the direction WK at the speed of two units of length, finds itself at point K, 800,000 km. from the Sun (Figure 88). At this distance the force of the Sun's attraction will, in a unit of time, impel the planet towards it 1.6 units of length; in the same interval it will move 2 units in the original direction of WK.

As a result it moves along the line KP -the diagonal of the parallelogram formed by the moves K1 and K2; this diagonal is equal to 3 units of length (Figure 88).

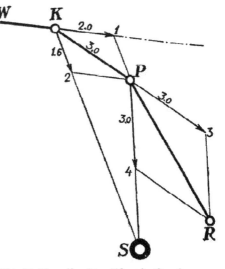

Upon reaching point P, the planet seeks to move farther along the direction KP with a speed of 3 units. But, under the influence of the Sun's attraction from the, distance SP = 5.8, it is impelled along the path P4=3 in the direction of SP.

Fig. 88. How the Sun S bends the planet's path WKPR.

As a result it will traverse the diagonal PR of the parallelogram.

There is no point in continuing to chart the path on this drawing -the scale is too big. Naturally, the smaller the scale the greater the part of the planet's path we can reproduce and the less will the acuteness of the, angles distort the likeness of the chart to the planet's true path.

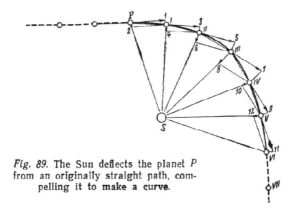

Fig. 89. The Sun deflects the planet *P* from an originally straight path, compelling it to make a curve.

Figure 89 reproduces the same picture on a smaller scale for our imagined encounter between the Sun and any celestial object similar in mass to the foregoing planet. It shows clearly how the Sun deflects the newcomer from its original path, making it follow the curve P—I—II—III—IV—V—VI. The angles here are not so sharp and we can easily link up the positions of the planet by drawing an even curve.

What is this curve? Geometry furnishes the answer. Place a sheet of tracing paper on the drawing and copy (Figure 89) 6 points of the planet's path chosen at random. Number the points (Figure 90) in any order and link them up in the same sequence by drawing straight lines. This gives a hexagram inscribed within the planet's path, partly with crossed sides. Now continue the straight line 1-2 to its intersection with the line 4-5 at point I. Obtain similarly the point II at the intersection of the straight lines 2-3 and 5-6 and then the point III at intersection of the lines 3-4 and 1-6. If the curve we are examining is a so-called "conic section," i.e., an ellipse, parabola, or hyperbola, the points I, II, III should be on the same straight line. This is the geometrical theorem (not among those studied at secondary school) known as "Pascal's Hexagram."

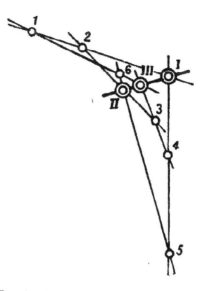

A careful drawing will always give the indicated points of intersection on one straight line. This proves that our curve is either an ellipse, parabola or hyperbola. The first apparently does not apply to Figure 89 since the curve is not locked, and so the planet moves according to a parabola or hyperbola. The ratio between the original velocity and force of attraction

Fig. 90. Geometrical proof of planetary motions about the Sun along conic sections (see text for details).

is such that the Sun can but deflect the planet from its straight path. It is powerless to compel the planet to revolve around itself, or, as astronomers would say to "capture" it.

Let us now elucidate the second law of planetary motion, the so-called law of areas. Take a good look at Figure 21. The 12 points divide it into 12 sections; although not of equal length, we know that the planet covers them in the same times. Linking points 1, 2, 3, etc., with the Sun, we obtain 12 figures which, if the points were joined by chords, would approximate to triangles. Measure their bases and heights and compute their areas. We would find that each triangle is equal in area. In other words, we arrive at Kepler's second law: *The radius-vectors of planetary orbits sweep equal areas in equal times.* Thus, in a way, the compass helps us to understand the first two laws of planetary motion. To elucidate the third law, we lay aside the compass, take up pen and paper and perform a few mathematical exercises.

When Planets Fall onto the Sun

Have you ever tried to imagine what would happen to the Earth if upon meeting an obstacle it were suddenly to cease its flight around the Sun? In the first place, naturally, the vast store of energy invested in it as a moving body would turn into heat and warm it up. And since it dashes along its orbit dozens of times faster than a bullet, one can easily imagine that when turned into heat the energy of its motion would kindle a monstrous conflagration that would immediately convert the world into an immense cloud of flaming gas...

But even if the sudden standstill did not produce this result, the Earth would, nevertheless, perish in flames; the attraction of the Sun would impel it headlong to die in the latter's fiery embrace.

This fateful fall-out would begin very slowly, literally at a snail's pace, with the Earth approaching the Sun only by 3 mm. in the first second. But each second would see a progressive increase of velocity until at the last second, it would reach 600 km. At this incredible speed the Earth would plunge into the Sun's flaming surface.

How long would the catastrophe take? How long would our doomed world writhe in agony? We can reckon the duration by applying Kepler's third law which governs the motions not only of the planets but also of comets and all celestial bodies in general moving in space under the influence of a central force of gravity. This law, which links the time of a planet's revolution (its "year") with its distance to the Sun, states:

The squares of the periods of revolution of the planets about the Sun are in the same ratio as the cubes of the major semi-axes of their orbits.

In the given case we can liken the globe in its sunward flight to an imaginary comet moving along a strongly extended flat ellipse the extreme points of which are located on the Earth's orbit and in the centre of the Sun. The major semi-axis of the orbit of such a comet would, apparently, be half the major semi-axis of the Earth's orbit. So let us find the period of revolution of our imaginary comet.

On the basis of Kepler's third law we form the ratio

(Earth's period of revolution)^2/(Comet's period of revolution)^2=(Major semi-axis of comet's orbit)^3/(Major semi-axis of Earth's orbit)^3

The period of the Earth's revolution is 365 days.

Taking the major semi-axis of its orbit as 1, the major semi-axis of the comet's orbit would be expressed by the fraction 0.5. Now our ratio will be:

$$365^2/(\text{Comet's period of revolution})^2 = 1/(0.5)^3$$

Hence

$$\text{Comet's period of revolution} = 365/\sqrt{8}$$

What we are interested in is not the full period of the revolution of our imaginary comet, but in half the period, or rather in the length of its flight one way from the Earth's orbit to the Sun, this being the time it will take the Earth to fall onto the Sun. Let us calculate it:

$$365/\sqrt{8} : 2 = 365/2\sqrt{8} = 365/5.65$$

Consequently, to find how long it will take the Earth to fall out on the Sun we should divide the year by $\sqrt{32}$ or by 5.65. In round number this is 65 days.

So we find that after the sudden standstill in orbital flight, the Earth would need more than two months to fall onto the Sun.

We easily see that this simple formula evolved on the basis of Kepler's third law can be applied not only to the Earth, but to all the planets, and even to all the satellites. In other words, to find how long it would take a planet or satellite to fall onto the primary, we divide the period of revolution by or 5.65.

Thus, it would take Mercury, the Sun's nearest planet with an 88-day period of revolution, 15.5 days to fall onto the Sun, Neptune, whose "year" is equal to 165 Earth years, would need 29 years, Pluto 44.

How long would it take the Moon to fall onto the Earth were it to come to a sudden standstill? We divide the Moon's period of revolution of 27.3 days by 5.6 and obtain almost five days exact. And not only the Moon, anybody the same distance away would fall onto the Earth in five days, provided its starting velocity is nil; in falling it would conform only to the action of the Earth's gravitation (for simplicity's sake we discount the Sun's influence).

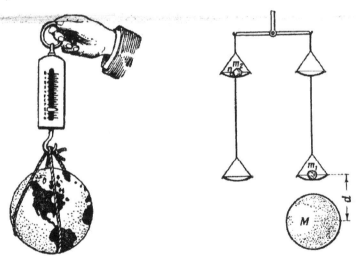

Fig. 91. What scales were
used to weigh the Earth?

Fig. 92. One way of ascertaining the Earth's mass; the Jolly balance.

By using the same formula we can easily establish how long it took Jules Verne's De la terre à la lune characters to get there.[3]

-3- The calculation is given in my book Travels Through Space

Vulcan's Forge

We shall use the rule we have evolved to solve a curious problem from the realm of mythology. The ancient Greek legend about Vulcan mentions casually how he once dropped his forge, its downward flight from the heavens to the Earth taking nine days. For the ancients this period fitted in with their notion of the measureless height of the abode of the gods; from the top of the Pyramid of Cheops the forge would have reached the Earth after five seconds!

It will not take us very long to find the universe of the ancient Greeks, if measured by this token, a rather modest one by modern concepts.

We know that it would take the Moon five days to reach the Earth; nine days were needed for the legendary forge. Hence the "sky" from which the forge, fell is farther than the Moon's orbit. But how much farther? Multiplying 9 days by $\sqrt{32}$ we find the forge's period of revolution around the Earth, that is supposing it to be a satellite of our planet: $9 \times 5.6 = 51$ days. Let us now apply to the Moon and to our imaginary forge-satellite Kepler's third law and write out the ratio:

$(Moon's \quad period \quad of \quad revolution)^2/(Forge's \quad period \quad of \quad revolution)^2 = (Moon's \quad distance)^3/(Forge's \quad distance)^3$

Evaluating the ratio we find:

$$27.3^2/51^2 = 380000^3/(Forge's \; distance)^3$$

The result is 580,000 km. How modest in the light of modern astronomy was the ancient Greeks' notion of the height of the sky! Only $1\frac{1}{2}$ times farther than the Moon! The universe of the ancients ended roughly where to our mind it just begins.

The Boundaries of the Solar System

Kepler's third law also enables us to calculate the distance to the boundaries of our solar system provided we take the farther-most points (aphelia) of cometary orbits as the extreme. We have discussed this earlier, so we confine ourselves here to the appropriate calculations. In chapter III we spoke about comets with an extremely long period of revolution as much as 776 years. Let us find the distance x to this comet's aphelion, knowing that the distance of its nearest point to the Sun (perihelion) is 1,800,000 km. Taking the Earth as our second body we write out the ratio:

$$776^2/1^2 = [1/2\ (x+1,800,000)]^3/150,000,000^3$$

hence

$$x = 25,318,000,000 \text{ km.}$$

We see, therefore, that these, comets travel 182 times farther from the Sun than the Earth and 4.5 times farther than Pluto, the most distant of known planets.

The Error in Jules Verne's Book

The imaginary Gallium comet, which Jules Verne took as the place of action for his *Hector Servadac*, made one full revolution around the Sun in exactly two years. The book also notes this comet's aphelion as being 820 million km away from the Sun. Although the perihelion is not given, we are entitled to assert, basing ourselves on two figures which we shall adduce, that no such comet exists in our solar system. We can demonstrate this with calculations based on Kepler's third law.

Let us designate the unknown distance to the perihelion as

x million km. The major axis of the comet's orbit would then be expressed by $x+820,000,000$ km. and the major semi-axis as $(x+820)/2$ million km.

Comparing the comet's period of revolution and distance with the Earth's period of revolution and distance, we get, on the basis of Kepler's law, the ratio:

$$2^2/1^2=(x+820)^3/(2^3 \times 150^3)$$

Whence

$$x = {}^- 343.$$

The negative result obtained for the comet's nearest distance to the Sun testifies to the incongruity of the problem's original data. In other words, a comet with such a brief period of revolution as two years could not travel the distance from the Sun Jules Verne gives in his book.

How Was the Earth Weighed?

Once upon a time, so the story goes, there was a naive chap who in studying astronomy was completely flabbergasted by the fact that astronomers knew the names of the stars. But, talking seriously, perhaps the most astounding achievement of the astronomers was their success in weighing both the Earth on which we live and the remote celestial luminaries. Indeed, how were the Earth and sky weighed and what scales were used?

First, how did they weigh the Earth? What do we mean by "the weight of the globe"? What we call weight is a body's pressure on its base, or its pull at a point of suspension. Neither of these can be applied to the Earth, since it rests on nothing and hangs on nothing. Consequently, from this point of view the Earth has no weight. What then did astronomers define when they weighed "the Earth"? They ascertained its mass. Indeed, when we ask the shop assistant to weigh us a kilogram of sugar we are not in the least interested in its pressure on its base, or its pull at a spring balance. As regards sugar, we are interested in something quite different. We want to know how many cups of sweet tea we can make, or, in other words, the quantity of sugar we get in this kilogram.

However, there is only one way to measure quantity. That is to find the measure of the Earth's attraction for a body. We take it for granted that equal quantities accord with equal masses, judging of the mass of a body only by the force of its attraction, since attraction is proportional to the mass.

Turning now to the Earth's weight, we say that we shall be able to find its "weight," when we know its mass; thus we must understand the task of weighing the Earth as that of ascertaining its mass.

Here is a description of one method used to solve this problem (Jolly's method, 1871.) Figure 92 shows us a very sensitive pair of balances, with two light, upper and lower scales, 20 to 25 cm. apart, suspended from each arm. We place a spherical weight with the mass m_1 on the- right lower scale. To retain equilibrium we place the weight m_2 on the- left upper scale. These weights are not equal; since they are at different heights the Earth attracts them with different forces. If we now place under the right lower scale a large lead ball with the mass M, equilibrium will be disturbed, as the mass M of the lead ball will attract the m_1 with the force F_1 which is proportional to the product of these- masses and inversely proportional to the square of the distance d separating their centers:

$$F = k\, m_1\, M/d^2$$

where k is the so-called constant of gravity.

To restore the disturbed equilibrium we place on the left upper scale- a small weight with the mass n. The force with which it presses against the scale is equal to its weight, i.e., is equal to the force of the-attraction of this weight by the entire mass of the Earth. This force F' is equal to

$$F' = k \; nm \; m_e/R^2$$

where m_e is the Earth's mass, and R its radius.

Discounting the negligible effect of the lead ball on the weights in the left upper scale, we can state the condition of equilibrium as follows:

$$F = F', \text{ or } m_I \; M/d^2 = nm \; m_e/R^2$$

In this proportion all the values, save, the Earth's mass m_e can be measured. So we find m_e. In the experiments just mentioned, M=5775.2 kg, R=6,366 km, d=56.86cm, m_I = 5.00 kg, and n=589 mgr.

As a result, the Earth's mass is 6.15 x 10^{27} gr.

The value of the Earth's mass based on extensive measurements yields m_e =5.974 x 10^{27} gr. or about 6,000 trillion tons. The margin of error is not more than 0.1 per cent. That is how the astronomers ascertained the mass of the Earth. We are fully entitled to say that they weighed the Earth, for whenever we weigh things on a pair of scales, we actually determine not the weight, not the force of their attraction by the Earth, but their mass; we merely establish that the mass of a body is equal to the mass of the weights.

What Is Inside the Earth?

Here it will be appropriate to point to an error met sometimes in popular science books and articles. To simplify things, the job of weighing the Earth is pictured like this: First, astronomers

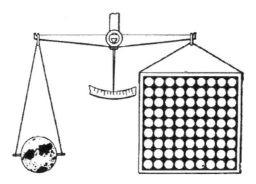

Fig. 93. The Earth "weighs" 81 times more than the Moon.

measured the mean weight of 1 cu. cm. of our planet, i.e., its specific weight and then, after geometrically reckoning its volume, weighed the Earth by multiplying specific weight by volume. The method, however, is not a feasible one; we cannot take a direct measurement of the Earth's specific weight, as only the comparatively thin outer crust[4] is accessible and we have no idea whatever of the much larger remaining part.

We already know that the problem was handled in a totally different way; the Earth's mass was ascertained before its mean density. This was found to be 5.5 grams per 1 cu. cm, much greater than the mean density of the rock of its crust. It can be inferred, then, that there are some very heavy substances in the bowels of the Earth. It was believed earlier, on the basis of surmised specific weight (and other factors) that the core of our plant consisted of iron, strongly compressed by the pressure of surrounding masses. It is now believed that, generally speaking, the central regions of the Earth do not differ in constitution from the crust, but that their density is greater due to the terrific pressures.

-4- The minerals in the Earth's crust have been investigated only to a depth of 25 km; according to estimates a mineralogical study has been made only of 1/83 of the Earth's volume.

Weighing the Sun and the Moon

Strange though it may seem, it proved incomparably simpler to ascertain the weight of the distant Sun than that of the much nearer Moon. (Naturally, with respect to these bodies we use the word "weight" in the same sense as for the Earth; here we mean ascertainment of their masses.)

The Sun's mass was found in the following manner. Experiments have shown that at a distance of 1 cm, 1 gr attracts 1 gr. with a force

Taking M as the Sun's mass (in grams), m as the Earth's mass, and D -the distance between them- equal to 150,000,000 km., their inter- attraction in milligrams will be:

On the other hand this force of attraction is the same centripetal force, which holds our planet to its orbit and which, according to the rules of mechanics, is equal (also in milligrams) to mV^2/D, where m is the Earth's mass (in grams), V its circular velocity equal to 30 km/sec=3,000,000 cm/sec, and D the distance from the Earth to the Sun. Hence,

This equation gives us the unknown M (expressed, as we have said, in grams):

$$M = 2 \times 10^{33} \text{ gr} = 2 \times 10^{27} m$$

Dividing this mass by the Earth's mass, i.e., $2 \times 10^{27}/6 \times 10^{21}$ we obtain 1/3 of a million.

There is another way of ascertaining the Sun's mass, based on

Kepler's third law. On the basis of the law of universal gravitation the third law is reduced to the following formula:

$$(M_s+m_1) T_1^2/(M_s+m_2) T_2^2 = a_1^3/a_2^3$$

where M_s is the Sun's mass, T the planets sidereal period of revolution, a the mean distance from the planet to the Sun and m the planet's mass. Applying this law to the Earth and the Moon, we obtain:

$$(M_s+m_e) T_e^2/(m_e+m_m) T_m^2 = a_e^3/a_m^3$$

After evaluating a_e, a_m, and T_e, T_m with the figures derived through observation, and disregarding in the numerator of the first approximation the Earth's mass, as minute in comparison with the Sun's mass, and in the denominator the Moon's mass, as minute in comparison with the Earth's mass, we obtain:

$$M_s/m_e = 300,000$$

Knowing the Earth's mass, we can determine the Sun's mass.

So, the Sun is a third of a million times heavier than the Earth.

We can easily reckon also the mean density of the solar sphere; to do this we need but divide its mass by its volume. We learn that the Sun's density is roughly a quarter of the Earth's density.

As for the Moon, "although nearer to us than all other celestial bodies," as one astronomer puts it, "it is harder to weigh than Neptune, the (then) most distant planet." The Moon has no satellite that would enable us

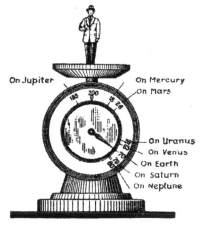

Fig. 94. How much would we weigh on the different planets.

to reckon its mass in the way that we have just found the Sun's mass. Astronomers had to resort to more complicated methods, of which I shall mention only one. It consists in comparing the high Sun-produced and lunar tides.

Tidal height depends on the mass and distance of the causal body. Since we know the Sun's mass and distance and the Moon's distance, we can ascertain the Moon's mass by comparing tidal heights. We shall return to this calculation when explaining the tides. Meanwhile here is the final result: the Moon's mass is 1/81 of the Earth's mass (Figure 93). Knowing the Moon's diameter, we calculate its volume which we find to be 1/49 of the Earth's volume. Hence, the mean density of our satellite is 49/81 = 0.6 of the Earth's density.

Consequently, the Moon consists generally of a more friable substance than the Earth, but with greater density than that of the Sun. We shall see in the table below that the Moon's mean density is greater than the mean densities of most planets.

Weight and Density of Planets and Stars

The method used to "weigh" the Sun can be used for weighing any planet with at least one satellite.

Knowing the mean velocity v of the satellite's orbital motion and its mean distance D from the planet, we can place the equal sign between the centripetal force holding the satellite to its orbit, mv^2/D, and the force of the inter-attraction between satellite and planet, i.e., kmM/D^2 where k is the force attraction, exerted by 1 gr on 1 gr, 1cm away, m is the satellite's mass and M the planet mass:

$$mv^2/D = kmM/D^2$$

Whence

$$M = Dv^2/k$$

a formula from which we can easily deduce the planet's mass M.

Kepler's third law is also applicable to this particular case:

$$(M_s + m_{planet}) \, T_{planet}^2 / (m_{planet} + m_{satellite}) \, T_{satellite}^2 = a_{planet}^3 / a_{satellite}^3$$

Here again, ignoring the bracketed minute members in the sums we obtain the ration of the Sun's mass to the planet's mass.

We can use the same method for binary stars, with the only difference that here we obtain not the separate masses of the stars in the pair but the sum of their masses.

The masses of Mercury and Venus were established by taking into account their perturbation upon each other, and upon the Earth, and also upon the motions of certain comets.

As for the asteroids, whose masses are so negligible that they do not noticeably perturb each other, ascertainment of their masses is, generally speaking, impossible. All we know -and even this is guesswork- is the limit of the aggregate masses of the pygmy planets.

Knowing a planet's mass and volume we can easily ascertain its mean density. Here are the results (Earth's density =1):

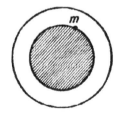
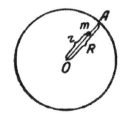

Fig. 95. A body inside a spherical casing has no weight.

Fig. 96. On what does the weight of a body inside a planet depend?

Fig. 97. A diagram showing the change in the weight of a body the nearer it approaches the centre of a planet.

Mercury	1.00	Jupiter	0.24
Venus	0.92	Saturn	0.13
Earth	1.00	Uranus	0.23
Mars	0.74	Neptune	0.22

We see that the Earth and Mercury rank first for density in our planetary family. The reason for the small mean density of the bigger planets is that the solid core of each is enveloped in a vast atmosphere, which, while of small mass, greatly increases the planet's apparent volume.

Weight on the Moon and on the Planets

Laymen with hazy notions of astronomy often wonder why scientists, having never been to either the Moon or planets, speak with such assurance about their surface gravity. In reality it is not very difficult to calculate how much a body transported to another world should weigh. All we need know is the radius and mass of the chosen celestial body.

Let us ascertain, say, the gravitational pull of the Moon. We already know that the Moon's mass is 1/81 of the Earth's mass. If the Earth had such a small mass its surface gravity would be but 1/81 of what it is now. But according to Newton's law a sphere's attraction is the same as if all its mass were concentrated in its centre. The Earth's centre is at the distance of its radius to the surface. Naturally the Moon's centre is the distance of its radius to the surface. However, the Moon's radius is 27/100 of the Earth's, and when distance is 100/27 times less, the force of attraction increases $(100/27)^2$ times. Consequently, the Moon's surface gravitation will be ~ $100^2/(27^2 \times 81)$ ~1/6 of the Earth's.

Thus a 1 kg weight would weigh only 1/6 kg on the Moon's surface.

Naturally, the loss in weight would be discovered only by using a spring balance (Figure 94), not a pair of scales.

Curiously enough, if the Moon had any water a swimmer there would feel exactly the same as he does on Earth. Although his weight would be six times less, the weight of the water displaced by him would be the same number of times less. Thus the proportion between them would be the same as on Earth, with the swimmer immersed in Moon water exactly in the same fashion as in Earth water.

But the effort to get out of the water would take much less energy on the Moon, because with the weight of the swimmer's body being less, the strain on his muscles would be less.

Here is a table of gravity values on different planets compared with the Earth's gravity.

On Mercury	0.26	On Saturn	1.13
On Venus	0.90	On Uranus	0.84
On Earth	1.00	On Neptune	1.14
On Mars	0.37	On Jupiter	2.64

The table places the Earth fourth for gravity in the solar system, after Jupiter, Neptune and Saturn.

Record Weight

Peak gravity is reached on the surface of such "white dwarfs" as Sirius B, mentioned in Chapter IV. We can easily imagine that the tremendous mass of these luminaries, with their comparatively small radius, lends them quite a significant surface gravity. We shall calculate this factor for one of the stars in the constellation of Cassiopeia, which has a mass 2.8 times that of the Sun and a radius half that of the Earth. Bear in mind that the Sun's mass is 330,000 times that of the Earth, and that the surface, gravity of the foregoing star will, therefore, be 2.8 x 330,000 x 2^2=3,700,000 times that of the Earth's.

On the surface of this star, 1 cm^3 of water, which weighs 1 gr on the surface of our Earth, would weigh nearly 3.75 tons! In this amazing world 1 cu. Cm of the matter making up the star (36,000,000 times denser than water) would have the monstrous weight of

$$3,700,000 \text{ x } 36,000,000= 133,200,000,000,000,000 \text{ gr.}$$

A thimbleful of substance weighing 100 million tons is something which only a short while ago was beyond the boldest flight of imagination.

Weight in the Depths of the Planet

To what extent would a body's weight change if placed far inside a planet, say at the bottom of a fantastically deep mine?

Many mistakenly think that a body at the bottom of such a mine as being nearer to the centre of the planet, to the point that attracts all bodies, would be heavier. This line of reasoning is wrong, however, because the force of attraction, far from increasing, decreases the deeper we go into a planet. The reader will find a comprehensible explanation of this phenomenon in my *Physics for Entertainment*. To avoid repetition I shall merely note the following.

Mechanics shows that a body placed in the cavity of a homogeneous spherical casing loses its weight completely (Figure 95). It follows, therefore, that a body inside a solid homogeneous sphere is attracted only by the matter contained in the sphere where the radius is equal to the distance of the body from the centre (Figure

96).

Using these propositions we can easily formulate a law according to which the weight of a body changes the nearer it gets to the centre of a planet. Designating the radius of the planet (Figure 97) as R and the distance of the body from its centre as r, we find that the gravity acting on the body at this point increases $(R/r)^2$ times and simultaneously decreases $(R/r)^3$ times (since the attracting part of the planet has decreased the said number of times). In the final analysis, the force of gravity would decrease

$$(R/r)^3/(R/r)^2, \text{ i.e., } R/r \text{ times.}$$

Consequently in the depths of a planet the weight of a body would decrease the number of times the distance to the centre has shortened. For planets the size of our Earth with its radius of 6,400 km, a body 3,200 km, deep would weigh half as much, and at 5,600 km deep 6,400/(6,400-5,600) = 8 times less. At the centre of the planet the body would lose its weight completely, since (6,400-6,400)/6,400 = 0.

Incidentally, this could have been foreseen without calculations, since a body in the centre of a planet is attracted by the surrounding medium with the same force from all sides.

What we have said applies to a planet with a homogeneous density, and can be applied to real planets only with certain reservations. In particular, speaking of the Earth the density of which in its depths is greater than near the surface, the law according to which gravitation changes in accordance with nearness to the centre, somewhat departs from this rule; to a certain (comparatively small) depth the gravitation increases, only deeper down does it begin to dwindle.

The Problem of the Steamer

Question

When is a steamer lighter, on a moonlit or on a moonless night?

Answer

The problem is knottier than it seems. We cannot say straight out that on a moonlit night a steamer and in general all objects on the moonlit half of the globe are lighter than on a moonless night because "they are attracted by the Moon." In attracting the steamer the Moon simultaneously attracts the entire Earth. In a void, under the effects of gravitation, all bodies move with the same velocities; the Moon's gravitation lends both Earth and steamer the same acceleration so we cannot discern any loss in weight. And yet the moonlit steamer is lighter than the vessel sailing along on a moonless night.

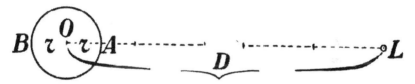

Fig. 98. The effect of lunar attraction on an Earth particle.

The reason is this. Suppose O on Figure 98 designates the centre of the Earth, and A and B the steamer at two diametrically opposed points of the globe, r is the Earth's radius, and D the distance from the Moon's centre L to the Earth's centre O. M is the Moon's mass and m the steamer's mass. To simplify the calculations we shall make points A and B coincide with the Moon's zenith and nadir. The Moon's force of attraction at point A (i.e., on a moonlit night) is

$$kMm/(D-r)^2$$

where $k=1/15,000,000$. At point B (on a moonless night), the

Moon attracts the streamer with a force equal to

$$kMm/(D+r)^2$$

The difference between the two attraction forces is equal to

$$kMm \times 4r/[D^3(1-(r/R)^2)^3]$$

Since $(r/R)^2=(1/60)^2$. It is an exceedingly negligible quantity, it is discounted. This greatly simplifies the expression and we get:

$$kMm \times 4r/D^3$$

Let us change it to:

$$(kMm/D^2) \times (4r/D) = (kMm/D^2) \times (1/15)$$

What is kMm/D^2? We easily guess that this is the force of attraction exerted by the Moon on the steamer at distance D from its centre. On the Moon's surface, a steamer with the mass m weighs $m/6$. At the distance D from the Moon it is attracted by the latter with the force $m/6D^2$. Since D is 220 lunar radii, hence

$$kMm/D^2 = m/6 \times 220^2 \approx m/300,000$$

Returning now to the difference of attractions we get

$$kMm/D^2 \times 1/15 \approx m/300,000 \times 1/15 = m/4,500,000$$

Assuming that the steamer weighs 45,000 tons, we find the difference in weight on a moonlit and moonless night to be

$$45,000,000/4,500,00 = 10 \text{ kg}$$

So while the difference is insignificant, the steamer is lighter on a moonlit night than on a moonless night.

Lunar and Solar Tides

The problem we have just examined will also help to explain the main reason for the high and low tides. It would be, wrong to think that the tidal wave rises simply because of the pull of the Sun or the Moon. We have explained that the Moon, in addition to attracting objects on the Earth's surface, attracts the Earth itself. The point, however, is that the centre of the Earth is farther from the attracting source than the particles of water on its surface facing the Moon. The corresponding difference in attraction is reckoned in the same way we reckoned the difference in the attractions acting on the steamer. At any point where the Moon is in the zenith every kilo of water is attracted $2kMr/D^3$ times stronger than every kilo of matter at the Earth's centre, while the attraction for each kilo of water at the diametrically opposed point is correspondingly weaker.

It is this difference that causes the water to rise in both cases above the surface of the Earth. In the first case it takes place because the water moves towards the Moon more than the solid part of the Earth does. In the second case, the Earth's solid part moves more towards the Moon than the water.[5]

The Sun's gravitation has a similar effect on ocean water. Which of the two is stronger, solar or lunar attraction? If we compare their direct gravitations we find that the Sun's is the stronger. The Sun's mass, as we know, is 330,000 times that of the Earth, while the Moon's mass is 81 times less, or 330,000 x 81 times less that of the Sun. The distance between the Sun and the Earth is 23,400 times the Earth's radius, while the distance between the Moon and the Earth is only 60 times the Earth's radius. Consequently, the ratio between the Sun's attraction for the Earth and the Moon's attraction is

$$330,000 \times 81/23,400^2 : 1/60^2 \approx 170$$

-5- We note here only the main reason for the high and low tides; actually this phenomenon is more complicated and is caused also by other reasons (the centrifugal effect of the Earth's rotation around the common centre of the masses of the Earth and the Moon, etc.).

This means that the Sun attracts every terrestrial object with a force 170 times greater than the Moon. One might think, therefore, that solar tides are higher than lunar tides. Actually we see the reverse; lunar tides are higher. This fully conforms to the formula $2kmM/D^3$. Designating the Sun's mass by M_s, the Moon's mass by M_m the distance to the Sun by D_s, and the distance to the Moon by D_m. we find the ratio between the tide-raising forces of the Sun and the Moon to be

$$2kM_s^r/D_s^3 : 2kM_m^r/D_m^3 = M_s/M_m \times D_m^3/D_s^3$$

We assume the Moon's mass to be known and equal to 1/81 of the Earth's mass.

Then, knowing that the Sun is 400 times farther away than the Moon we find

$$M_s/M_m \times D_m^3/D_s^3 = 330{,}000 \times 81 \times 1/400^3 = 0.42$$

Hence, the solar tides should be $2^1/2$ times lower than lunar tides.

Here it will be in place to note how, by comparing the heights of lunar and solar tides, the Moon's mass was ascertained. The height of one or the other tide cannot be observed separately as the Sun and the Moon always act jointly. But we can measure tidal height when the two luminaries act in conjunction (i.e., when the, Moon and the Sun are situated on a straight line with the Earth), and when they act contrary to each other (when the straight line joining the Sun and the Earth is perpendicular to the straight line joining the Moon and the Earth). Observations showed the second tide to be 0.42 of the height of the first. Designating the Moon's tide-raising force by x and that of the Sun by y we get

$$(x+y)/(x-y) = 100/42$$

Whence

$$x/y = 71/29$$

Consequently, applying the foregoing formula we obtain

$$(M_s/M_m) \times (D_m/D_s)^3 = 29/71$$

Or

$$(M_s/M_m) \times (1/64{,}000{,}000) = 29/71$$

Since the Sun's mass $M_s = 330{,}000\, M_e$ where M_e is the Earth's mass, from the last equality we easily find

$$M_e/M_m = 80,$$

i.e., the Moon's mass is 1/80 that of the Earth. A more accurate reckoning gives the Moon's mass as 0.0123 (of the Earth's mass).

The Moon and Weather

Many are interested in the effect of the high and low lunar tides in our planet's ocean of air on atmospheric pressure. This question has a long history. The tides in the Earth's atmosphere were discovered by the famous Russian scientist Lomonosov, who called them air waves. Although many have busied themselves with these air waves, nevertheless, distorted notions of their action are widely current. Laymen presume that the Moon causes huge tidal waves in the Earth's light and mobile atmosphere. Hence the conviction that the tides largely alter atmospheric pressure and are decisive in meteorology.

Such is not the case. We can prove theoretically that the height of an atmospheric tide cannot exceed that of the tide in the ocean. This assertion comes as a surprise; one would think that since the air, even in its lower denser layers, is well nigh a thousand times lighter than water, why should not the Moon's attraction cause it to rise a thousand times higher? This, how-ever, is no more of a paradox than the equal rapidity with which heavy and light bodies fall in a void.

Let us recall one of our schoolday experiments when the ball of lead in an empty tube fell no faster than the feather. The tide is the end result of the Earth's fall in space together with its lighter casings, under the effect of the gravitation of the Moon (and the Sun). In the cosmic void all bodies, the light and the heavy, fall with the same speed covering the same distance owing to gravi-tation. Provided, of course, that their distance from the centre of gravitation is the same.

From what has been said you will realize that the height of the atmospheric tides is the same as those in the ocean, off shore. Indeed if we address ourselves to the formula used to calculate tidal height, we shall find that it contains only the masses of the Moon and the Earth, the radius of the Earth and the distance from the Earth to the Moon. It calls for neither the density of the liquid raised nor the depth of the ocean. Even if we were to substitute air for water the result of the calculations would not be altered. We obtain the same height for an atmospheric tide as for an ocean tide. The latter, incidentally, is insignificant. Theoretically, the highest ocean tide is around half a meter and only the contours of shores and bottoms, by containing the tidal wave, raise it to 10m and more in some places. There are unique mechanisms for predicting the height of the tide at a given place and time from data on the position of the Sun and the Moon.

However, in the boundless ocean of air nothing can interfere with the theoretical picture, of the lunar tide and change its theo-retical peak height of half a meter. Such an insignificant rise can have but only the slightest effect on atmospheric pressure.

Laplace, when investigating the theory of air tides, concluded that the fluctuations of atmospheric pressure they caused would not exceed 0.6 mm. of the column of mercury, while the velocity of the wind produced would not exceed 7.5 cm/sec.

It is quite plain that atmospheric tides cannot play any essential role in weather making.

These considerations debunk all attempts of the "Moon seers" to forecast the weather by the position of the Moon in the skies.

Index

A

Achilles 140
Agamemnon 140
Algol 159
Altair 162
Amundsen 19, 20
aphelia 208
Arcturus 162
Around the World in Eighty Days 62

B

Bering Straits 61

C

Canopus 162, 164
Canopy of Black 90
Cape Dezhnev 61
Capella 159, 162
Cape of Good Hope 11, 18
Cape Verde Islands 62
Capricorn 28
Cassiopeia constellation 184
Centauri 187, 189
Compass 31

D

De la terre à la lune 206
Deimos 144
Deneb 191
Density of Planets 215
diurnal luminary 94
Dorado constellation 169
Double Planet 71

E

Earth's Axis 44
Earth's Path 49
Edgar Allan Poe 62

ISBN : 978-2917260180

36782477R00133

Made in the USA
Lexington, KY
05 November 2014